Mechanical Properties of Layered Rock Mass

AF840448

Xianhui Feng · Bin Gong

Mechanical Properties of Layered Rock Mass

Springer

Xianhui Feng
University of Science and Technology
Beijing
Beijing, China

Bin Gong
Brunel University of London
London, UK

ISBN 978-981-96-5141-2 ISBN 978-981-96-5142-9 (eBook)
https://doi.org/10.1007/978-981-96-5142-9

© The Editor(s) (if applicable) and The Author(s), under exclusive license to Springer Nature Singapore Pte Ltd. 2025

This work is subject to copyright. All rights are solely and exclusively licensed by the Publisher, whether the whole or part of the material is concerned, specifically the rights of translation, reprinting, reuse of illustrations, recitation, broadcasting, reproduction on microfilms or in any other physical way, and transmission or information storage and retrieval, electronic adaptation, computer software, or by similar or dissimilar methodology now known or hereafter developed.
The use of general descriptive names, registered names, trademarks, service marks, etc. in this publication does not imply, even in the absence of a specific statement, that such names are exempt from the relevant protective laws and regulations and therefore free for general use.
The publisher, the authors and the editors are safe to assume that the advice and information in this book are believed to be true and accurate at the date of publication. Neither the publisher nor the authors or the editors give a warranty, expressed or implied, with respect to the material contained herein or for any errors or omissions that may have been made. The publisher remains neutral with regard to jurisdictional claims in published maps and institutional affiliations.

This Springer imprint is published by the registered company Springer Nature Singapore Pte Ltd.
The registered company address is: 152 Beach Road, #21-01/04 Gateway East, Singapore 189721, Singapore

If disposing of this product, please recycle the paper.

Competing Interests The authors have no competing interests to declare that are relevant to the content of this manuscript.

Contents

Chapter 1
Introduction

Abstract Investigating the mechanical properties of layered rock mass is essential for ensuring the safety, stability, and efficiency of engineering projects such as tunnels, slopes, dams, and foundations. Layered rock masses often exhibit complex behaviors due to the presence of bedding planes, joints, and varying material compositions. These features significantly influence how the rock mass responds to stress, deformation, and failure. Understanding mechanical properties helps engineers predict how the rock will behave under different loading conditions. This is particularly important in layered formations, where anisotropy can lead to unexpected failures if not properly accounted for. Furthermore, accurate characterization of these properties supports the design of effective support systems and excavation methods, reducing construction risks and costs. It also aids in assessing long-term stability, especially in regions prone to seismic activity or weathering. In geotechnical engineering, neglecting the mechanical behavior of layered rock masses can result in catastrophic failures, environmental damage, and financial loss. Therefore, this book is not only a scientific necessity but also a practical requirement for sustainable and safe infrastructure development.

Keywords Rock mass · Bedding · Anisotropy · SHPB · Mechanical properties

With the development of human society, the demand for energy resources is increasing day by day. The development and utilization of tunnels, slopes and foundations have attracted growing attention from countries around the world. Currently, in the process of rapid mining of mineral resources, many shallow mineral resources are about to or have already faced depletion. At the same time, due to the limitations of marine resource mining technology, the exploitation of deep mine resources has gradually become inevitable trend for sustainable development.

Layered rock masses are generally sedimentary rocks, parametamorphic rocks and volcanic rocks with layered structures. The most obvious feature is the mature layered structure, including striped, laminated and rhythmic structures. From a macro perspective, it can be seen that the mineral content of rocks, the color ratio of rocks, the particle sizes of minerals, and structural structures show a certain sequence of

© The Author(s), under exclusive license to Springer Nature Singapore Pte Ltd. 2025

X. Feng and B. Gong, *Mechanical Properties of Layered Rock Mass*,
https://doi.org/10.1007/978-981-96-5142-9_1

reproducibility in the vertical direction. The periodic changes in mineral crystallization and accumulation fully display the characteristics of accumulation rock. Due to the existence of preferred structural planes, layered rock masses usually have transversely isotropic characteristics, that is, in the direction parallel to the dominant structural planes, the compositions and mechanical properties of rock mass are basically the same, and it is usually assumed to be isotropic media materials; In the direction perpendicular to the dominant structural planes, the mechanical properties are obviously different, with orthorhombic anisotropic characteristics, and the failure mechanism is also more complex. The deformation and strength of layered rock masses are often controlled by the alternation of bedding planes or weak planes, and are also restricted by the geological environment in which the rock mass is located (i.e., the seepage field, stress field, temperature field, humidity, etc.) and the influence of external loads (earthquake, ground stress, blasting, etc.).

There are increasing examples of large-scale infrastructure projects such as tunnels, slopes, shale gas development and dams being constructed in layered rock masses. Furthermore, in 2002, a landslide accident occurred in the Emeishan limestone mine of Sichuan Jinding Group due to deep hole blasting in the layered rock mass; in 2009, for the main plant and downstream main transformer of the Jinping Level 1 Hydropower Station in the southwest mountain canyon area, the increase in the measured displacement at its arch waist is caused by the unloading splitting and buckling failure of layered rock mass under high stress conditions (Wei et al. 2010); in addition, in the Wujiang River Basin in Chongqing, China, Wanzhou, western Hubei and other regions, large-scale bedding landslides controlled by weak interlayers have also occurred many times, seriously threatening people's property and life safety (Zhang 2015). The Gezhouba Project in China was suspended several times due to weak interlayers in the layered rock mass. Supplementary surveys and redesigns were carried out, and the project investment increased by more than ten times. In March 2010, a water leakage accident happened in the No. 16 coal seam excavation working face during construction at Luotuoshan Coal Mine in Wuhai, Inner Mongolia, China. The accident caused 32 deaths and a direct economic loss of 48.53 million Chinese yuan.

Many geotechnical projects, such as engineering blasting, protection engineering, and seismic exploration, involve the propagation and attenuation of stress in layered rock masses, as well as the deformation and damage of layered rock masses under load. The anisotropy of layered rock mass structure determines that its mechanical properties, stress wave propagation and attenuation characteristics are significantly different from those of single rock mass, and its formation is also more intricate. Based on the core issue of structural instability of layered rock mass, this book introduces our recent research results, outlines a series of studies on this issue from both physical tests and numerical simulations, and analyzes the evolution and development trends of layered rock mass fracture, with a view to providing reference for understanding the fracture and instability of layered rock masses.

1.1 Anisotropic Characteristics of Mechanical Properties of Layered Rock Mass

Layered rock mass is a complex natural ore body. Different rock masses contain different mineral compositions and various internal defects (joints, bedding, pores, cracks, etc.). Therefore, they exhibit homogeneity, anisotropy and transverse isotropy, etc. (Li et al. 2018) in mechanical properties. These characteristics have substantial impacts on the mechanical behavior of rock, including deformation capacity, strength, fracture toughness, and failure mode, and play an important decisive role in the structural stability of rock under various loads (Xu et al. 2019; Zhang et al. 2019).

The relevent research on the anisotropic characteristics of rock mass structures began in the middle of the last century. The main research content at that time was indoor experiments on the anisotropic strength and deformation characteristics of intact or directional jointed fracture rock materials. Segraves (1963) derived the general equation of anisotropic body elasticity theory based on the generalized Hooke law of continuum media, thus providing a theoretical basis for anisotropic theoretical research. Pinto (1966) obtained an orthotropic elastic modulus expression for a composite material composed of alternating layers of isotropic elastic materials with two different properties, which was used to study the fracture modes and failure conditions of schistose and bedding rocks. Gerrard (1982) represented the layered rock mass as equivalent uniform orthogonal elastic material and proposed the application of orthogonal layer theory to calculate and predict stress distribution. Singh (1973) used the equivalent medium method to obtain the elastic modulus expression of anisotropic rock mass and study the deformation parameters and mechanical properties of layered rock mass. Ramamurthy (1993) looked into the influence of the angle between bedding plane and main load axis on the deformation modulus and strength parameters of rock mass, analyzed its induced and inherent forms, and summarized the varying rules of anisotropic characteristics. Talesnick and Bloch-Friedman (1999) verified the effectiveness of the anisotropic deformation parameters of the three devices through conducting uniaxial compression, radial splitting and ring tests on three different anisotropic materials. Affected by the cementing strength of bedding planes and strong and weak bedding planes, the cracking time, cracking position and expansion direction of shale samples under loads are different, and the fracture modes present diversified characteristics. As the bedding angle increases during the uniaxial compression of shale samples, the failure mode of shale samples changes from vertical tension failure to shear slip failure along the weak bedding plane, and finally evolves into shear failure across bedding planes. The uniaxial compressive strength of shale samples shows a U-shaped change pattern that first decreases and then increases as bedding inclination increases. The strength change trend of shale samples is closely related to its fracture mode. When the bedding angle is horizontal and perpendicular to the loading direction, the specimen has the highest strength, and the strength is mainly controlled by the rock matrix.

In view of the current situation of such rock mass, many scholars have conducted multiple experiments on the influence of weak bedding plane and joints on the mechanical properties of rock mass. Due to the preferential orientation of mineral particles during the formation of sedimentary rocks, shale has obvious bedding structural characteristics, causing the tensile strength of shale to be greatly affected by bedding planes (Zeng et al. 2019). Because of the influence of beddings and joints on the compressive strength of rocks, there is a hidden danger of instability in the construction of geotechnical engineering. In order to control the stability of rock mass, it is necessary to clarify the influencing factors and relevant mechanical parameters of layered rock mass so that corresponding measures can be taken. In addition to anisotropic mechanical strength and final failure morphology, the fracture process and failure mode have always been another focus of researchers' observation and study (Huang et al. 2020). Zhao et al. (2022) utilized the asymmetric semi-circular bend (ASCB) method to conduct fracture tests on coal samples with different bedding angles between the loading direction and the bedding plane under Type II loading, and explored the effect of bedding angle on the fracture toughness and fracture mode of coal under Type II loading. Claesson and Bohloli (2002) classified the Brazilian splitting failure modes of anisotropic rocks into five categories, including tensile failure across and along bedding planes, shear failure across and along bedding planes and mixed failure. Xie et al. (2021) conducted uniaxial compression experiments and found that as the bedding angle θ increases, the failure mode of shale samples changes from tension failure penetrating the bedding plane to shear failure first, and then to splitting-shear mixed failure, and the shale bedding plane with weak cementation will be damaged before the matrix. Sang et al. (2015) viewed the anisotropic mechanical properties and failure characteristics of shale through confining pressure tests, and proposed two modes of shale failure, namely shear failure occurring inside the rock and sliding failure along the bedding plane. Besides, they optimized the drilling plan and fracturing design. According to the dynamic and static combined strength characteristics of bedding rock mass, Feng et al. (2022) raised a strength criterion for bedding rock mass that uniformly describes the transition from quasi-static to dynamic, through two parameters with clear physical meaning and easy to determine, namely, critical strain rate and bedding angle, which can effectively predict the strength of bedding rock mass under impact load within a certain range.

1.2 Constitutive Model of Layered Rock Mass

The stress of rock mass changes with strain increasing (Szwedzicki 2007). The purpose of rock strength criteria is mainly to reflect the failure mechanism of rocks through the relationship equations between relevant parameters (Aubertin et al. 2000). Currently, in the field of rock mechanics, the commonly used criteria are the maximum normal stress strength criterion, Mohr–Coulomb strength criterion, Hoek–Brown strength criterion, Griffith strength criterion, etc. (Jing 2003; Ming et al. 2013; Ozgur and Tugrul 2013; Peng 1975; Stephen and Thomas 1981). Because the layered rock mass has obvious transverse isotropy, the strength is not only related to the matrix of rock itself, but also has a great correlation with the structural plane and the angle of joint plane (Zhang et al. 2014).

In order to describe the strength characteristics of layered rock masses, many scholars have implemented a lot of research and work from theoretical and experimental aspects (Jaeger 1960; Juang et al. 2006; Yong et al. 2000), and put forward constitutive models for layered rock masses such as equivalent anisotropic constitutive model, Cossert constitutive model, elastic–plastic constitutive model, etc. The Goodman joint constitutive model has promoted the in-depth development of joints and bedding structural planes in studying anisotropic mechanical properties of joints or layered rock masses. Based on the Coulomb-Navier criterion, Jaeger (1960) gave the corresponding failure criteria for the slide fracture of rock masses containing single weak planes or multi-stratified weak surfaces. Tien et al. (2006) proposed the transverse isotropic rock failure criterion defined by seven material parameters based on two different failure modes: ① failure caused by sliding along the structural plane; ② non-sliding mode controlled by the rock matrix. Huang et al. (2010) came up with the hardening–softening model for layered rock mass composites to describe the anisotropy and impact fracture characteristics of such rock mass in terms of strength and deformation, and successfully applied the model to slope engineering and cavern engineering. Yin (2020) offered a new criterion for predicting the Brazilian splitting failure behavior of layered rocks based on triaxial compression, Brazilian splitting tests, triaxial cyclic loading and unloading tests, and unloading confining pressure tests. Fan et al. (2013) developed the nonlinear dynamic equivalent continuum method of rock mass and proved the superiority of equivalent viscoelastic continuum method in characterizing the dynamic characteristics of rock mass. Liu et al. (2023) took the coupled dynamic and static Hopkinson impact device as a platform to carry out dynamic Brazilian splitting tests and impact compression tests. They systematically summarized the tensile/compressive strength rules of bedding rock mass from static to dynamic under coupled dynamic and static stress and revealed the anisotropic mechanical response and instability mechanism of the impact resistance characteristics of bedding rocks under dynamic action.

1.3 Numerical Simulation Study of Layered Rock Mass

The layered rock mass structure has the characteristics of anisotropy, inhomogeneity, discontinuity and others, which is important for understanding the deformation and strength of bedding planes and weak interlayers in the rock mass. As a complement to laboratory experiments, numerical simulations have proven to be a powerful tool for in-depth understanding of anisotropic and heterogeneous material failure mechanisms of bedding rock masses. At present, common numerical calculation methods mainly include finite element method (FEM) (Li et al. 2020; Wang et al. 2011; Zhao et al. 2020), discrete element method (DEM) (Tomac and Gutierrez 2017), boundary element method (BEM) (Pan et al. 1998), discontinuous deformation analysis (DDA) (Wu et al. 2004), finite difference method (FDM) (Wang et al. 2014), spectral element method (SEM) (Peirce et al. 1992), etc., which analyze the safety and stability of layered rock mass under load from the perspective of the stress field and displacement field of the rock mass.

For example, Mao (2006) combined the Cosserat medium theory and the layered rock mass theory to establish the constitutive relationship of slate and investigated the fracture mode of slate based on single weak plane theory and Mohr-Coulumb criterion, so as to determine the varying patterns of maximum compressive strength with the angle of structural plan. Fan (2004) introduced the two-dimensional anisotropic damage constitutive model into the boundary element method and embedded the Dam-twofs program on the basis of the anisotropic Asio-twofs program to study various engineering problems of anisotropic fractured rock mass and anisotropic damaged rock masses with lining structure. Li (2013) through PFC^{2D}/FLA^{3D} revealed that the basic failure mode of horizontal layered jointed rock mass in high slopes under earthquake dynamics can be characterized by the sliding failure along the direction of earthquake source, and the penetrating failure surface is approximately arc-shaped. He (2021) built a PILE unit fracture simulation numerical model through the secondary development of FISH language to research the tensile, shear and tensile-shear composite fractures of anchor rods caused by large deformation of tunnel surrounding rock, and discussed the fracture process of arch rod under various loads, providing a reference for the structural design of tunnel anchor rods. Xia (2009) selected assembly language VC++6.0 to implement the algorithm of transversely isotropic elastic–plastic constitutive relations and embedded it in the $FLAC^{3D}$ finite difference numerical analysis software to realize numerical operations. Then, the in-situ stress of layered rock mass in tunnel site area has been inverted and studied, which offers reliable scientific basis for the design and construction of similar projects. Xia et al. (2016) used particle flow to establish a numerical model to study the effects of bonding strength, friction and occlusion on the compressive strength of layered rock mass from a mesoscopic perspective. Wu (2018) adopted the Largrange algorithm and Euler algorithm in the ANSYS-DYNA software to construct a constitutive model corresponding to brittle materials under high strain rates to simulate layered slate, and the stress wave time history curve and failure mode obtained were consistent with experimental results. With the aid of the dynamic analysis mode in $FLAC^{3D}$ software, some researchers simulated the stress, displacement distribution and change trend of rock samples when they were subjected to combined dynamic and static loads. For example, Du (2018) established an anisotropic model of coal considering bedding and cleat through $FLAC^{3D}$, analyzed the influence of bedding and other factors on the nonlinear mechanical behavior of impact-prone coal samples during the compaction stage and revealed the anisotropy of nonlinear behavior during the compaction stage. Duan et al. (2015) employed the discrete element method to simulate the indirect tensile strength, stiffness and fracture mode of layered rock mass Brazilian disc. This numerical model can be further applied to study the anisotropic properties of rocks like fracture toughness and strength criteria under different stress conditions. Wang et al. (2019a, b) selected discrete element PFC^{2D} to establish a discrete fracture network (DFN) model with different joint distributions, and quantitatively analyzed the changing trend and spatial anisotropy of tensile strength for layered rock mass. Wu et al. (2020) utilized RFPA^{2D}-DIP to map the geometry of calcite to a finite element mesh of shale mesoscale structure and analyzed the anisotropy of compressive strength and the fractal dimension of cracks. Meng et al. (2021) simulated the influence of the number and strength of bedding on the fracture mode and crack distribution of the specimen through the viscous zone model in ABAQUS. Wang et al. (2019a, b) modified the Tien-Kuo criterion by adjusting four parameters: maximum principal stress, minimum principal stress, bedding angle, and friction coefficient between beddings, and used the ANSYS-workbench statics module to conduct numerical simulation on multi-scale layered pages, which verified the improved rock failure criterion. Zhang et al. (2009) simulated the joint distribution and crack generation and expansion in rock mass by popular numerical methods, after which drew the conclusion that the failure modes of horizontal and vertical jointed rock masses are largely affected by joint distribution. Yang et al. (2019) chose red sandstone and gray sandstone with large differences in wave impedance to splice into layered composite rock mass samples, and then combined the DIC test system to capture the failure process. In their study, the dynamic characteristics and failure characteristics of composite rock mass was discussed and a comparative study was conducted on the differences in the composite rock mass when the stress wave went from hard to soft and from soft to hard. Li et al. (2014) explored the relationship between bedding angle, rock material and stress distribution law based on elastic-brittle-plastic constitutive with the help of three-dimensional finite difference software FLAC. Based on the improved Hamilton principle, Chen et al. (1981) studied the changing law of the interface crack tip dynamic profit intensity factor of a double-layer medium under dynamic impact load. Wünsche et al. (2007) employed the boundary finite element method combined with the convolution integral method aroused by the time domain discretization method to explore the anisotropy of crack shape under impact load and the impact of the load on the dynamic strength factor and stress intensity factor. Wang et al. (2023) adopted the discrete element bond contact model to investigate the compressive strength and failure mode of layered biotite granulite with different bedding angles. The study found that the single axial compressive strength of layered rock mass varies from gradual decrease to gradual increase as the layered angle increases, showing a U-shaped distribution. Liang et al. (2005) developed the rock fracture process numerical simulation software (RFPA) based on mesoscopic damage mechanics to study the fracture mechanism of layered rock samples under impact load, and discussed the crack fracture modes and failure criteria of transverse isotropic rock mass such as bedding angle and strength. Feng et al. (2022) through RFPA^{2D}, and deeply illustrated the anisotropy of layered rock mass under dynamic impact stress from the perspective of static and dynamic superposition of the background stress field that induces fine crack expansion.

1.4 Existing Problems and Development Trends

(1) There have been relatively mature studies on the static mechanical characteristics of layered rock mass. However, for deep mining in mine areas, the rock mass will also be subject to the static load of high in-suit stress. As a result, the research on the mechanical properties of layered rock mass under the dynamic-static combination load which is similar to deep mining is still not comprehensive enough. So far, the dynamic characteristics and fracture mechanism of rock under high stress state that are fully suitable for actual situation have not been obtained.

(2) At present, it is difficult to have a theoretical model or numerical model that simultaneously considers the mechanical properties and crushing characteristics of shale under various complex loads using various natural and technological factors. In the absence of physical experiments and numerical simulation studies, we can proceed from the two main lines of natural conditions and construction conditions to carry out multiple numerical simulations on factors like bedding inclination, strain rate, and in-situ stress, so that provide references for the rock burst and the pillar stability caused by rock blasting vibration under high stress in deep underground mining.

(3) In the field of rock mechanics, a lot of research has been done on the crushing mechanism and crushing characteristics of rock under a single static load or dynamic load. Meanwhile, the current research on rock crushing is mainly focused on uniaxial compression, direct tension, dynamic loading and shear, etc., while the research on coupled dynamic and static loading is relatively lacking, especially physical experimental research on rock crushing characteristics under coupled dynamic and static loading modes. Therefore, it is of great theoretical significance and engineering application value to carry out numerical simulation and experimental research on rock crushing under combined dynamic and static modes, so as to enrich the theory of rock crushing and develop new drilling and supporting equipment.

References

Aubertin, M., Li, L., Simon, R.: A multiaxial stress criterion for short- and long-term strength of isotropic rock media. Int. J. Rock Mech. Min. Sci. **37**(8) (2000)

Chen, W.H., Wu, C.W.: On elastodynamic fracture mechanics analysis of bi-material structures using finite element method. Eng. Fract. Mech. **15**(1–2), 155–168 (1981)

Claesson, J., Bohloli, B.: Brazilian test: stress field and tensile strength of anisotropic rocks using an analytical solution. Int. J. Rock Mech. Min. Sci. **39**(8), 991–1004 (2002)

Du, W.: Research on Mechanical Response and Energy Development Characteristics of Impact-Prone Coal. China University of Mining and Technology, Beijing (2018)

Duan, K., Kwok, C.Y.: Discrete element modeling of anisotropic rock under Brazilian test conditions. Int. J. Rock Mech. Min. Sci. **78**, 46–56 (2015)

Fan, Z.: Application of boundary element method on mining subsidence in stratified rock. J. China Coal Soc. **02**, 150–154 (2004)

Fan, L.F., Wong, L.N.Y.: Stress wave transmission across a filled joint with different loading/ unloading behavior. Int. J. Rock Mech. Min. Sci. **60**, 227–234 (2013)

Feng, X., Gong, B., Cheng, X., Zhang, H., Tang, C.A.: Anisotropy and microcrack-induced failure precursor of shales under dynamic splitting. Geomat. Nat. Haz. Risk **13**(1), 2864–2889 (2022)

Gerrard, C.M.: Equivalent elastic moduli of a rock mass consisting of orthorhombic layers. Int. J. Rock Mech. Min. Sci. Geomech. Abs. **19**(1), 9–14 (1982)

He, L.: Implementation and application of tension-shear fracture of anchor bolts in FLAC3D. Mod. Tunnel. Technol. **58**(03), 60–69 (2021)

Huang, S., Xu, J., Ding, X., et al.: Study of layered rock mass composite model based on characteristics of structural plane and its application. Chin. J. Rock Mech. Eng. **29**(04), 743–756 (2010)

Huang, B., Li, L., Tan, Y., Hu, R., Li, X.: Investigating the meso-mechanical anisotropy and fracture surface roughness of continental shale. J. Geophys. Res. Solid Earth **125**(8) (2020)

Jaeger, J.C.: Shear failure of anistropic rocks. Geol. Mag. **97**(1), 65–72 (1960)

Jing, L.: A review of techniques, advances and outstanding issues in numerical modelling for rock mechanics and rock engineering. Int. J. Rock Mech. Min. Sci. **40**(3) (2003)

Juang, Y.M., Kuo, M.C., Juang, C.H.: An experimental investigation of the failure mechanism of simulated transversely isotropic rocks. Int. J. Rock Mech. Min. Sci. (2006)

Li, X.: Research of the Seismic-Induced Failure Mechanism of Layered Rock High Slope. China University of Geosciences (2013)

Li, A., Shao, G., Fan, H., et al.: Investigation of mechanical properties of soft and hard interbedded composite rock mass based on meso-level heterogeneity. Chin. J. Rock Mech. Eng. **33**(S1), 3042–3049 (2014)

Li, A., Xu, N., Dai, F., Gu, G., Hu, Z., Liu, Y.: Stability analysis and failure mechanism of the steeply inclined bedded rock masses surrounding a large underground opening. Tunnel. Undergr. Space Technol. **77**, 45–58 (2018)

Li, Z., Tao, M., Du, K., Cao, W., Wu, C.: Dynamic stress state around shallow-buried cavity under transient P wave loads in different conditions. Tunn. Undergr. Space Technol. **97**, 103228 (2020)

Liang, Z., Tang, C., Li, H., et al.: Numerical study on failure process of transversely isotropic rock subjected to uniaxial compression. Rock Soil Mech. **01**, 57–62 (2005)

Liu, H., Feng, X., Liu, L., Li, T., Tang, C.: Mechanical properties and failure characteristics of anisotropic shale with circular hole under combined dynamic and static loading. Int. J. Rock Mech. Min. Sci. **170**, 105524 (2023)

Mao, H.: Experiment Studies and Theoretical Analysis of Water-Weakening Properties of Slates. The Chinese Academy of Science (2006)

Meng, Y., Jing, H., Liu, X., Yin, Q., Wei, X.: Experimental and numerical investigation on the effects of bedding plane properties on the mechanical and acoustic emission characteristics of sandy mudstone. Eng. Fract. Mech. **245**, 107582 (2021)

Ming, T., Xibing, L., Diyuan, L.: Rock failure induced by dynamic unloading under 3D stress state. Theoret. Appl. Fract. Mech. **65** (2013)

Ozgur, Y., Tugrul, U.: Three dimensional numerical rock damage analysis under blasting load. Tunnel. Undergr. Space Technol. Incorp. Trench. Technol. Res. **38** (2013)

Pan, E., Amadei, B., Kim, Y.: 2-D BEM Analysis of Anisotropic Half-Plane Problems—Application to Rock Mechanics (1998)

Peirce, A.P., Spottiswoode, S., Napier, J.A.L.: The spectral boundary element method: a new window on boundary elements in rock mechanics. Int. J. Rock Mech. Min. Sci. Geomech. Abs. **29**(4), 379–400 (1992)

Peng, S.S.: A note on the fracture propagation and time-dependent behavior of rocks in uniaxial tension. Int. J. Rock Mech. Min. Sci. Geomech. Abs. **12**(4) (1975)

Pinto, J.L.: Stresses and Strains in an Anisotropic-Orthotropic Body. Proc. First Congress of the International Society for Rock Mechanics, Lisbon (1966)

Ramamurthy, T.: Strength and modulus responses of anisotropic rocks. In: Hudson, J.A. (ed.) Comprehensive Rock Engineering, vol. 1, pp. 313–329, Pergamon (1993)
Sang, Y., Yang, S., Zhao, F., Hou, B.: Research on anisotropy and failure characteristics of Southern Marine Shale Rock. Drill. Product. Technol. **38**(02), 71–74+10 (2015)
Segraves, W.A.: In: Lekhnitskii, S.G. (ed.) Theory of Elasticity of an Anisotropic Elastic Body, p. 404. Diagrams, 6 × 9 in., San Francisco, Holden-Day, Inc.. Price, $10.95 (1963)
Singh, B.: Continuum characterization of jointed rock masses: part I—the constitutive equations. Int. J. Rock Mech. Min. Sci. Geomech. Abs. **10**(4), 311–335 (1973)
Stephen, N.C., Thomas, J.A.: Dynamic tensile strength of lunar rock types. J. Geophys. Res. Solid Earth **86**(B3) (1981)
Szwedzicki, T.: A hypothesis on modes of failure of rock samples tested in uniaxial compression. Rock Mech. Rock Eng. **40**(1) (2007)
Talesnick, M.L., Bloch-Friedman, E.A.: Compatibility of different methodologies for the determination of elastic parameters of intact anisotropic rocks. Int. J. Rock Mech. Min. Sci. **36**(7), 919–940 (1999)
Tien, Y.M., Kuo, M.C., Juang, C.H.: An experimental investigation of the failure mechanism of simulated transversely isotropic rocks. Int. J. Rock Mech. Min. Sci. **43**(8), 1163–1181 (2006)
Tomac, I., Gutierrez, M.: Coupled hydro-thermo-mechanical modeling of hydraulic fracturing in quasi-brittle rocks using BPM-DEM. J. Rock Mech. Geotech. Eng. (2017)
Wang, S.Y., Sloan, S.W., Liu, H.Y., Tang, C.A.: Numerical simulation of the rock fragmentation process induced by two drill bits subjected to static and dynamic (impact) loading. Rock Mech. Rock Eng. **44**(3), 317–332 (2011)
Wang, T., Yang, C., Yan, X., Li, Y., Li, J.: Dynamic response of underground gas storage salt cavern under seismic loads. Tunn. Undergr. Space Technol. **43**(7), 241–252 (2014)
Wang, D.J., Tang, H., Shen, P., Su, X., Huang, L.: Co-effects of bedding planes and parallel flaws on fracture evolution in anisotropic rocks. Eng. Geol. **264**, 105382 (2019a)
Wang, Q., Li, Y., Yang, W., Jiang, Z., Liu, D.: Finite element simulation of multi-scale bedding fractures in tight sandstone oil reservoir. Energies **13**(1), 131 (2019b)
Wang, P., Ma, C., Liu, C., et al.: Anisotropic behavior of excavated layered rock mass subjected to compression considering the joint roughness. Bull. Eng. Geol. Env. **82**, 373 (2023)
Wei, J., Deng, J., Wang, D., et al.: Characterization of deformation and fracture for rock mass in underground powerhouse of Jinping I hydropower station. Chin. J. Rock Mech. Eng. **29**(06), 1198–1205 (2010)
Wu, L.: Investigation on Law and Application of Damage Evolution for Layered Rock Mass Subjected to Impact Loading. Wuhan University of Technology (2018)
Wu, J.H., Ohnishi, Y., Nishiyama, S.: Simulation of the mechanical behavior of inclined jointed rock masses during tunnel construction using Discontinuous Deformation Analysis (DDA). Int. J. Rock Mech. Min. Sci. (2004)
Wu, Z., Lou, Y., Yin, S., Wang, A., Liu, H., Sun, W., Zuo, Y., Chen, B.: Acoustic and fractal analyses of the mechanical properties and fracture modes of bedding-containing shale under different seepage pressures. Energy Sci. Eng. **8**(10), 3638–3656 (2020)
Wünsche, M., Zhang, C., Sladek, J., Sladek, V., Hirose, S.: Interface crack in anisotropic solids under impact loading. Key Eng. Mater. **348–349**, 73–76 (2007)
Xia, B.: Experimental Study on the Mechanism of Instability of Layered Rock Mass in Deep Buried Tunnel. Chongqing University (2009)
Xia, L., Zeng, Y., Zhang, S.: Influence of meso-mechanical parameters of bedding plane on strength characteristics of layered rock mass. J. Yangtze River Sci. Res. Inst. **33**(07), 68–75 (2016)
Xie, J., Lu, H., Chen, L., et al.: Micro scopic heterogeneity and mechanical anisotropy of the laminated shale in Long-maxi Formation. Bull. Geol. Sci. Technol. **40**(3):67–77 (2021)
Xu, G., He, C., Yan, J., Ma, G.: A new transversely isotropic nonlinear creep model for layered phyllite and its application. Bull. Eng. Geol. Environ. (2019)
Yang, R., Li, W., Fang, S., et al.: Experimental study on impact dynamic characteristics of layered composite rocks. Chin. J. Rock Mech. Eng. **38**(09), 1747–1757 (2019)

Yin, P.: Study on Mechanical Behavior and Hydraulic Fracturing Mechanism of Longmaxi Formation Shale in Southern Sichuan Basin. China University of Mining and Technology (2020)

Yong, M.T., Tsao, P.F.: Preparation and mechanical properties of artificial transversely isotropic rock. Int. J. Rock Mech. Min. Sci. **37**(6), 1001–1012 (2000)

Zeng, J., Liu, J., Zhang, Y.: Influence of bedding effect on tensile strength of black shale and its energy analysis. High Speed Railw. Technol. **10**(2), 38–43 (2019)

Zhang, S.: Limestone Bedding Landslide Formation Mechanism and Stability Research in Three Gorges Reservoir Wujiang River Valley. Chengdu University of Technology (2015)

Zhang, Q.B., Zhao, J.: A review of dynamic experimental techniques and mechanical behaviour of rock materials. Rock Mech. Rock Eng. **47**(4), 1411–1478 (2014)

Zhang, L., Lu, S., Liu, H.: Mechanism analysis and numerical simulation of the blast failure modes of jointed rock masses. Metalmine **397**(07), 16–19 (2009)

Zhang, Z., Jiang, D., Liu, W., Chen, J., Li, E., Fan, J., Xie, K.: Study on the mechanism of roof collapse and leakage of horizontal cavern in thinly bedded salt rocks. Environ. Earth Sci. **78**(10), 292 (2019)

Zhao, R., Tao, M., Zhao, H., Cao, W., Li, X., Wang, S.: Dynamics fracture characteristics of cylindrically-bored granodiorite rocks under different hole size and initial stress state. Theoret. Appl. Fract. Mech. **109**, 102702 (2020)

Zhao, Y., Sun, Z., Gao, Y., Wang, X., Song, H.: Influence of bedding planes on fracture characteristics of coal under mode II loading. Theoret. Appl. Fract. Mech. **117**, 103131 (2022)

Chapter 2
Failure Characteristics of Shales Under Impact Brazilian Splitting

Abstract The mechanical behaviors of shale exhibiting transversely isotropic characteristics under dynamic loading hold immense significance for addressing structural instability in geotechnical engineering. To unravel the impact of transverse isotropy on the deformability and tensile strength of shale subjected to dynamic loads, a series of impact Brazilian tests were conducted on shale specimens using the split Hopkinson pressure bar (SHPB) testing system. High-speed digital image correlation technology was utilized to monitor the fracture propagation process meticulously. The experimental results reveal a notable anisotropy in failure strength, with variations observed based on the bedding angle of the embedded layers. Specifically, shales with vertical beddings typically exhibit higher tensile strengths compared to those with parallel beddings. The predominant failure mode observed is the interplay between tensile and/or shear fractures. As the loading rate increases, layer-activated fractures become more prevalent. Furthermore, four distinctive failure patterns of transversely isotropic shales, characterized by different mechanisms, were identified under dynamic Brazilian testing. The deformation and fracture response of shales under dynamic loading were found to be highly sensitive to the strain rate. Consequently, a novel strength criterion was proposed, incorporating the strain rate effect, based on the Brazilian splitting data and a dynamic coordinate system. This criterion not only accurately represents the dynamic strength behavior but also offers fresh insights into the influence of the strain rate on strength. It has proven its effectiveness in predicting the dynamic strength characteristics of shale.

Keywords Shale · SHPB test · Anisotropy · Mechanical properties · Failure characteristics

2.1 Introduction

Due to long-term geological activities, rock masses in nature commonly exhibit anisotropic features such as foliations, fabrics, and bedding planes (Chong and Boresi 1990; Liang et al. 2019; Chen et al. 2022; Yu et al. 2022; Feng et al. 2022a). This

© The Author(s), under exclusive license to Springer Nature Singapore Pte Ltd. 2025
X. Feng and B. Gong, *Mechanical Properties of Layered Rock Mass*,
https://doi.org/10.1007/978-981-96-5142-9_2

anisotropy significantly influences rock mechanical behaviors, including deformability, strength, fracture toughness, and failure patterns, thereby playing a crucial role in ensuring the safety of rock structures subjected to various loadings (Debecker and Vervoort 2009; Liang et al. 2014; Cho et al. 2012; Gong et al. 2019a; Wang et al. 2022a). Furthermore, dynamic loadings such as blasting, impact, and earthquakes are frequently encountered in infrastructure construction and mining activities. The dynamic response of rock masses (Tavallali and Vervoort 2010; Zhu et al. 2021, 2022), particularly those with transverse isotropy, is largely governed by their structural features.

In recent years, numerous experimental and theoretical studies have been conducted on the mechanical anisotropic behavior of bedded shale (Dong et al. 2019; Yang et al. 2019; Raef et al. 2018). These investigations have primarily focused on conventional dynamic uniaxial compression and direct shear tests (Heng et al. 2015). Mahanta et al. (2018) analyzed bearing capacity, deformation, and fracture attributes and mechanisms at various strain rates through uniaxial compressive testing. Yang et al. (2020) discussed the effect of bedding on the mechanical parameters of shale rocks using confining pressure tests. Liu et al. (2015) suggested that the strengths of coal rocks with varying bedding directions exhibit significant differences, indicating anisotropy under both static and dynamic uniaxial compression. To achieve substantial theoretical advancements and practical applications, research on the loading rate of bedding rock fragmentation should not be confined to the macroscopic scale (Zhang et al. 2000). Therefore, it is imperative to further understand static and dynamic fractures of rock at the microscopic scale (Ma et al. 2018; Zhao et al. 2021). Zhang et al. (2000) monitored crack propagation near the fractured surface of dynamic rock samples, while Niu et al. (2021) discussed the evolution of microscopic damage and the dynamic response of sandstones subjected to freeze–thaw cycles. However, the dynamic failure mechanisms of transversely isotropic shales subjected to varying impact loadings have not been fully elucidated, and the impact of the loading rate on the bearing characteristics of transversely isotropic shales at the microscale remains unclear.

In the field of experimental rock mechanics, two-dimensional (2D) digital image correlation (DIC) has increasingly emerged as a powerful tool for studying crack propagation in rocks (Feng et al. 2022b; Gong et al. 2022; Wang et al. 2022b, c). Studies have proven that the formation of bedding surfaces and microcracks significantly affects the mechanical behaviors and crack growth of rocks (Szwedzicki 2006; Gong et al. 2019b; Jing 2003). Pan et al. (2021) investigated the influence of inner bedded planes in front of the artificial crack tip on crack evolution related to bedding planes. Nath and Mokhtari (2018) applied optical techniques to measure strain on the surface of shale samples with various mineralization angles through indirect tensile experiments. Zhou et al. (2021) used centrally straight-notched Brazilian discs of shales to study crack initiation and development at various load angles using the DIC method. Cho et al. (2012) analyzed the strong deformation and bearing anisotropy of rocks, including Asan gneiss, Boryeong shale, and Yeoncheon schist. Identifying different fracture types is crucial for analyzing the strength mechanism of anisotropic rock (Debecker and Vervoort 2009; Ramamurthy 1993). Wu et al.

(2021) demonstrated that fractures propagate less along bedding interfaces as the strain rate increases. Qiu et al. (2017) found that the length of the main fracture propagating along bedding planes in Brazilian disc tests decreases as the dip angle increases under dynamic loading. However, the effects of the loading rate, bedding angles, and crack development on indirect dynamic tensile strengths and final failure modes remain unclear.

Characterizing the tensile strengths of shales is significant for numerous geotechnical and geophysical projects (Gholami and Rasouli 2013; Pufahl et al. 2014). Due to the difficulty of direct tension, the Brazilian disc test has been proposed as a convenient alternative method for obtaining the tensile strengths of rocks (Perras and Diederichs 2014). The tensile strengths of shales are related to temperature (Qiu et al. 2017), loading rate (Wu et al. 2021), degree of anisotropy (Cho et al. 2012), and other factors. Many theoretical analyses have also been conducted to predict the strength anisotropy of shales from both experimental and theoretical perspectives, such as those by Nova and Zaninetti (1990), Hobbs (1967), Lee and Pietruszczak (2015), and Jaeger (2009). Hobbs (1967) predicted the tensile strengths of rocks using Griffith crack theory. Jaeger (2009) proposed the single weakness plane theory. However, additional dynamic loading may cause potential adverse consequences (Yang et al. 2020; Wu et al. 2021; Zhao 2000). Predicting excavation-induced damage by drill-and-blast operations is crucial for assessing the long-term stability and functionality of tunnels, caves, repositories, etc. To achieve this goal, it is necessary to understand the dynamic mechanical behaviors of rock masses, such as shales. However, determining the constitutive model and strength index of rock materials for this purpose must consider the strain rate effect.

In the present study, Brazilian splitting tests were conducted using shale specimens with seven bedding angles to ascertain crack propagation and investigate fracture mechanisms under impact loads using the split Hopkinson pressure bar (SHPB) testing instrument. The progressive fracture process was captured using DIC technology, which computed the complete displacement field of each sample disc. Scanning electron microscopy (SEM) technology was then used to detect the microscopic morphology of rock fractures. Through systematic experimental research, the strength characteristics and failure mechanisms of shales subjected to various strain rates were discussed at both the microscale and macroscale. Furthermore, based on the Nova-Zaninetti fracture pressure model, a new strength criterion considering the strain rate effect has been proposed according to the Brazilian splitting tests. The established strength criterion can effectively represent the dynamic strength law and provide new insights into the strain rate effect on shales.

2.2 Experimental Setup and Method

2.2.1 Sample Preparation

In this study, shale samples exhibiting a low degree of metamorphism, sourced from Lushan City in Jiangxi Province, China, were meticulously selected. To explore the impact of anisotropic characteristics on the mechanical behavior of shale rocks, seven distinct groups of shale specimens were prepared, each with bedding angles θ of 0°, 15°, 30°, 45°, 60°, 75°, and 90°, respectively. These specimens featured a diameter of 75 mm and a thickness of 30 mm. The bedding angle θ signifies the angle formed between the orientation of the layers and the loading direction, which was indicated by two mutually perpendicular dashed lines marked on the specimens. Adhering to the guidelines set by the International Society for Rock Mechanics and Rock Engineering (Aminzadeh et al. 2019) and the German Geotechnical Society (Lepique 2008), the surfaces of the samples were meticulously polished.

To gain insights into the mineral composition and bedding surface characteristics of the tested shale samples, as depicted in Fig. 2.1a, mineral analysis was conducted using X-ray Diffraction (XRD) and Scanning Electron Microscopy (SEM). The lithostratigraphic protolith of the Lushan copper ore deposit primarily comprises acid igneous rocks, exhibiting a lack of maturity in both rock structure and composition. These rocks contain numerous unstable clasts, including rock fragments, feldspar, and iron-magnesium minerals, indicative of limited transportation, minimal weathering, and weak alteration. Matrix cement, which serves as a particle support, is the predominant cement type. The source rock formed during the sedimentary syndepositional period, undergoing relatively mild compaction. Tectonic forces led to localized stress variations, creating oblique symmetrical pressure shadows within the rocks. Mild fracturing, alteration, and mineralization processes are evident within the rock matrix.

The shale samples exhibit transverse isotropy, characterized by alternating light and dark layers with nearly parallel boundaries. Darker layers are associated with

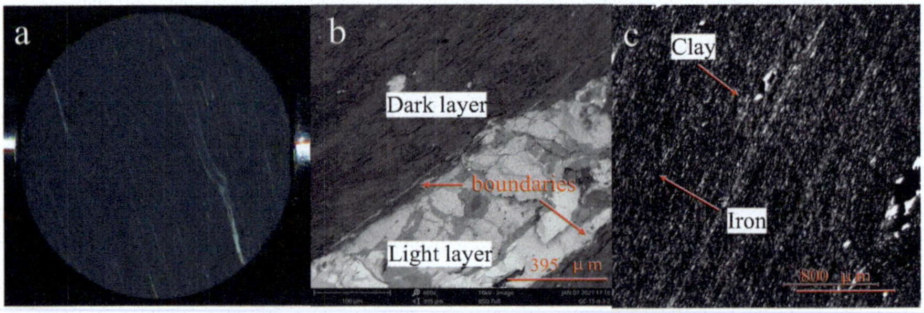

Fig. 2.1 Microscopic structure of the shale sample. Modified after Feng et al. (2024)

finer grain textures, while lighter layers correspond to coarse grain textures, as illustrated in Fig. 2.1b. The mineral composition of the investigated sections, as shown in Fig. 2.1c, includes clay (ranging from 60 to 65%), silty debris (30–35%), iron (5%), and carbonaceous matter (1–3%). Furthermore, the light layers display a disordered distribution and are predominantly composed of coarse-grained textures.

2.2.2 *Testing Apparatus and Methodology*

In the current study, the Split Hopkinson Pressure Bar (SHPB) testing system emerged as an invaluable tool in the laboratory for assessing the dynamic mechanical behaviors of rocks. The system comprised a gas gun, a striker, an incident bar, and a transmitted bar. Specifically, the striker measured 200 mm in length, while the incident and transmitted bars were 3000 mm and 1800 mm long, respectively, all sharing a uniform diameter of 75 mm. The shale samples under investigation were securely clamped between the incident and transmitted bars. To monitor the incident, reflected, and transmitted wave signals, two strain gauges were attached to both the incident and transmitted bars. Data acquisition was facilitated by an 8-channel DL750 Scope Corder digital oscilloscope and a 6-channel dynamic strain amplifier. Additionally, a Photron SA-Z high-speed digital camera, capable of capturing at a rate of 100,000 frames per second with a resolution of up to 640 × 280 pixels, was employed to document the shale failure process during impact. Notably, the striker and bars were crafted from hardened 60 Si2Mn material, exhibiting a Young's modulus of 211 GPa, a density of 7850 kg/m^3, and a P-wave velocity of 5102 m/s.

To quantitatively analyze the deformation behavior of the shale specimens, this study utilized the Digital Image Correlation (DIC) technique. DIC offers an optical, numerical, full-field measurement capable of capturing intricate, inhomogeneous deformation fields within the tested samples. By comparing images captured by the digital camera at various loading stages, changes in the deformation characteristics on the shale samples' surface could be accurately determined. Prior to speckle spraying, the samples' surface was thoroughly cleaned, with a white matte paint serving as the background and randomly distributed black spots comprising the speckle pattern.

The processed SHPB data encompassed incident, reflected, and transmitted strain histories recorded by the strain gauges adhered to the bars. Furthermore, a three-wave analysis approach was employed to compute the strain and stress characteristics of the shale samples, following methodologies outlined by Lifshitz and Leber (1994) and Sharafisafa et al. (2020). Utilizing one-dimensional elastic wave theory, the relationship between load and time at the contact ends was derived.

$$P_1(t) = EA_\mathrm{e}[\varepsilon_I(t) + \varepsilon_R(t)] \tag{2.1}$$

$$P_2(t) = EA_e\varepsilon_T(t) \tag{2.2}$$

The average load $P(t)$ at both ends of the sample is:

$$P(t) = \frac{P_1(t)) + P_2(t)}{2} = EA_e \frac{\varepsilon_I(t) + \varepsilon_R(t) + \varepsilon_T(t)}{2} \tag{2.3}$$

According to the assumption of stress uniformity, the stress balance at both ends of the sample is as follows:

$$\varepsilon_I(t) + \varepsilon_R(t) = \varepsilon_T(t) \tag{2.4}$$

Bring Eq. (2.3) into Eqs. (2.1) and (2.2) to get

$$u_s = \int_0^t C_0(\varepsilon_I - \varepsilon_R)dt = -2\int_0^t C_0\varepsilon_R dt \tag{2.5}$$

$$P(t) = EA_e\varepsilon_T(t) \tag{2.6}$$

where P_1 and P_2 represent the loading forces at the interfaces between the rock samples and the incident and transmitted bars, respectively, ε_I, ε_R, and ε_T denote the incident, reflected, and transmitted strain waves, respectively, monitored by the strain gauges attached to the bars. E_b and C_0 stand for the Young's Modulus and P-wave velocity of the bar material, respectively. L refers to the length of the rock sample, and A_b corresponds to the cross-sectional area of the bar.

Equations (2.5) and (2.6) are utilized to calculate the strain and force experienced by the dynamic tensile test sample within the SHPB setup. Based on the static Brazilian disc splitting principle, the tensile strength of the sample can be expressed as follows:

$$\sigma_t = \frac{2P}{\pi Dt} \tag{2.7}$$

where P is the failure load, D is the specimen diameter, and t is the specimen thickness.

Because the fractures of layer shale are not originated from the center of the rock disc, so this formula was only used to calculated the nominal dynamic tensile strength in this study. By combining Eqs. (2.6) and (2.7), the indirect tensile strength of the dynamic nominal tensile sample can be obtained:

$$\sigma_t = \frac{2P(t)_{\max}}{\pi DL} = \frac{2EA_e\varepsilon_T(t)_{\max}}{\pi DL} \tag{2.8}$$

where L is the thickness of the sample and D is the diameter; A_e is the cross-sectional area of the incident bar and the transmission bar; E is the elastic modulus of the incident bar and the transmission bar.

2.3 Experimental Results

2.3.1 *Effect of Bedding Angle*

To investigate the impact of bedding angle on rock fracture patterns, seven distinct bedding angles ranging from 0° to 90° were chosen for this study. Figure 2.2 showcases the corresponding cracking processes of the shale samples, with the strain rate maintained approximately constant at 40 s^{-1}. To accurately depict the characteristics of crack initiation and propagation, the photographs captured by a high-speed camera were binarized, which unfortunately obscured the bedding features. The results revealed that, regardless of the bedding angle, the failure process of bedded shales under impact can be categorized into three typical stages: crack initiation, crack propagation, and complete failure.

In the first stage, crack initiation occurs between 120 and 140 μs after impact. Notably, cracks did not originate at the center of the disc. Instead, microscopic cracks were first observed at both ends of the disc, particularly when $\theta = 90°$. Additionally, the orientation of these initial cracks was strongly influenced by the bedding angle of the shale samples, except when $\theta = 90°$.

During the crack propagation stage, which occurs from 160 to 190 μs after impact, the initial cracks continue to propagate and secondary cracks begin to form. Notably, the orientation of these secondary cracks varies with the bedding angle. Specifically, at $\theta = 0°$, 15°, and 30°, the bedding plane plays a dominant role in the propagation of secondary cracks, guiding them along the bedding interfaces. When $\theta = 45°$, 60°, and 75°, the secondary cracks exhibit a curved shape and their propagation is influenced by both the axial force and the bedding plane, meaning they propagate within the angle defined by the bedding plane and the axial loading direction. Conversely, at $\theta = 90°$, the bedding interface has a negligible impact on the failure mode, and the disc splits into two semicircles along the loading direction.

In the complete failure stage, the initial and secondary cracks further propagate and coalesce, leading to the emergence of a significant number of microcracks. This coalescence ultimately results in the overall splitting failure of the shale samples.

2.3.2 *Effect of Strain Rate*

This section systematically examines the impact of strain rate on the failure patterns of shale samples with varying bedding angles. By adjusting the velocity of the striker, a range of strain rates from 30 to 320 s^{-1} was considered, and the corresponding results are presented in Fig. 2.3. The analysis reveals that the strain rate has a significant influence on the failure patterns of shale specimens, regardless of their bedding angle. Specifically, as the strain rate increases, the failure severity of the shale samples intensifies. Additionally, the influence of bedding planes on rock fracture patterns becomes more pronounced at higher strain rates. At relatively low strain rates (e.g.,

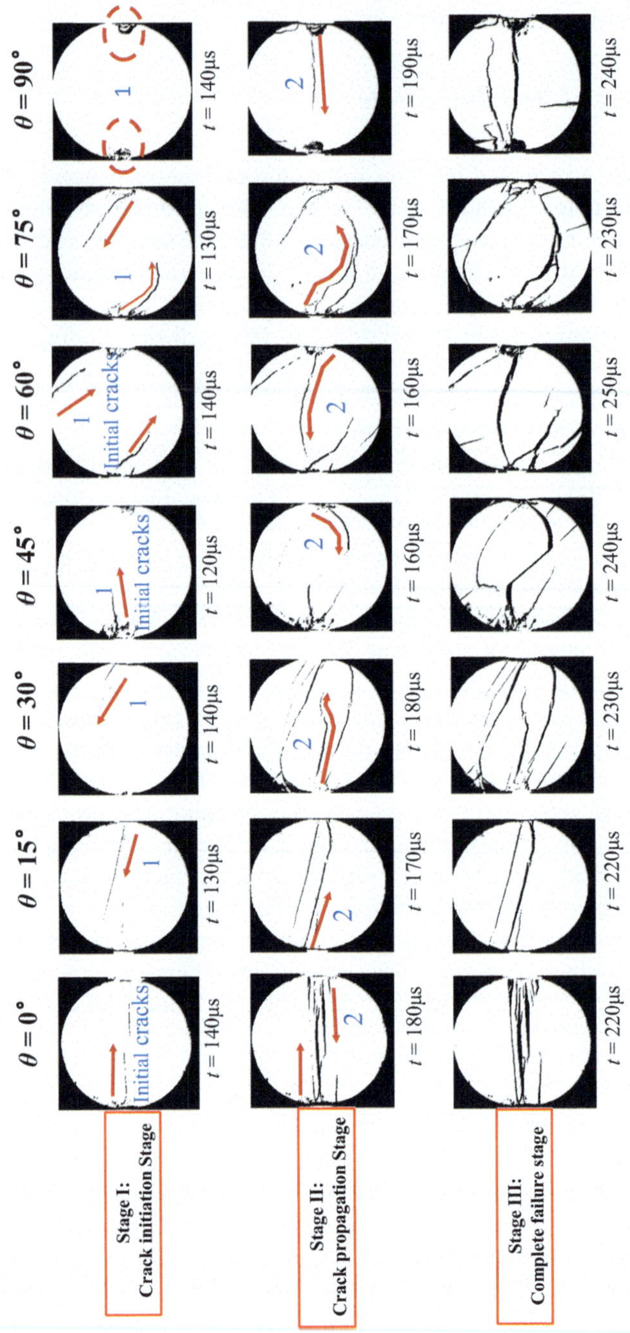

Fig. 2.2 Failure patterns of the shale samples with different bedding angles. Modified after Feng et al. (2024)

$\dot{\varepsilon} = 30\ \mathrm{s}^{-1}$), splitting failure can be observed. However, at higher strain rates (particularly $\dot{\varepsilon} = 200\ \mathrm{s}^{-1}$ and $\dot{\varepsilon} = 300\ \mathrm{s}^{-1}$), the bedding plane plays a dominant role in the failure of the shale samples, guiding crack propagation along the rock stratification plane. Notably, for $\theta = 90°$, splitting failure occurs consistently across all strain rate variations.

2.3.3 Dynamic Nominal Tensile Strength and Energy Absorption

Figure 2.4 illustrates the nominal tensile strengths of layered slates with varying bedding angles. The data reveals that the measured dynamic nominal tensile strengths exhibit distinct patterns of change for different bedding angles. Notably, these strengths increase rapidly as the strain rate escalates from 30 to 320 s^{-1}.

During the ongoing process of rock deformation and failure, the storage and consumption of energy occur concurrently with loading. To gain a more accurate understanding of rock deformation and failure, it is advantageous to approach rock mechanics from an energetic perspective. To quantify the influence of both bedding angle and strain rate on the dynamic fracture of shales, a systematic analysis of their energy absorption characteristics was conducted. The energy absorption ($\mathrm{W_c}$) during the dynamic fracture of the rock can be calculated using the following formula:

$$\mathrm{W}_c = W_i - \mathrm{W}_r - \mathrm{W}_t \tag{2.9}$$

where W_i, W_r and W_t refer to the energy of the incident wave, the reflected wave and the transmitted wave, which could be determined by:

$$W = A_b E_b C \int \varepsilon(t)^2 dt \tag{2.10}$$

where A_b, E_b and C stand for the cross-sectional area, the Young's modulus & the wave velocity of the bars. $\varepsilon(t)$ denotes the variation of the strain measured by strain gauges during the SHPB tests, which could be the incident strain wave, the reflected strain wave or the transmitted strain wave.

Figure 2.5 displays the energy consumption of shales subjected to various strain rates. The data indicates that, regardless of the strain rate, the energy consumed by shale samples initially decreases but then increases as the bedding angle rises. The energy consumption reaches a minimum at bedding angles of 30° or 60° and peaks at bedding angles of 0° or 90°. By comparing the energy consumption values with the dynamic strength characteristics depicted in Fig. 2.5, it is evident that the energy consumed by samples with different bedding angles also exhibits a strain rate correlation effect, with energy consumption increasing as the strain rate level rises. Furthermore, the energy absorbed by the rock sample can be categorized into

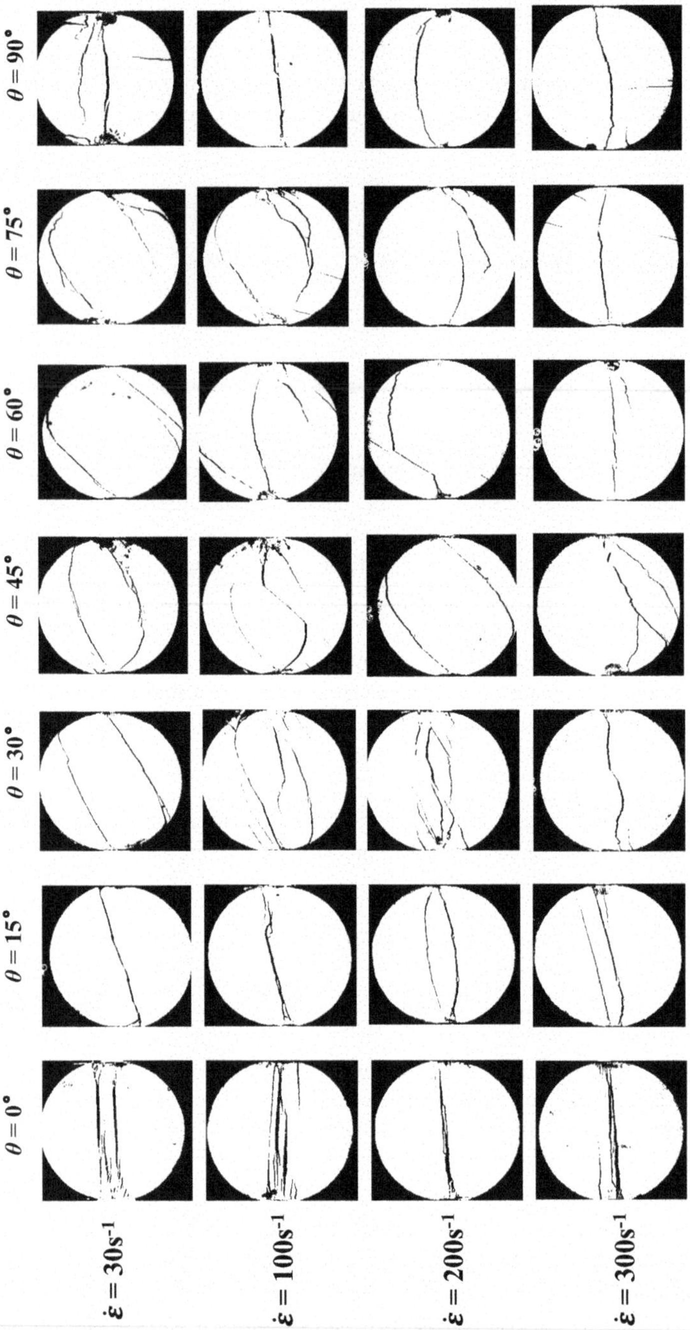

Fig. 2.3 Failure patterns of the specimens with different bedding angles. Modified after Feng et al. (2024)

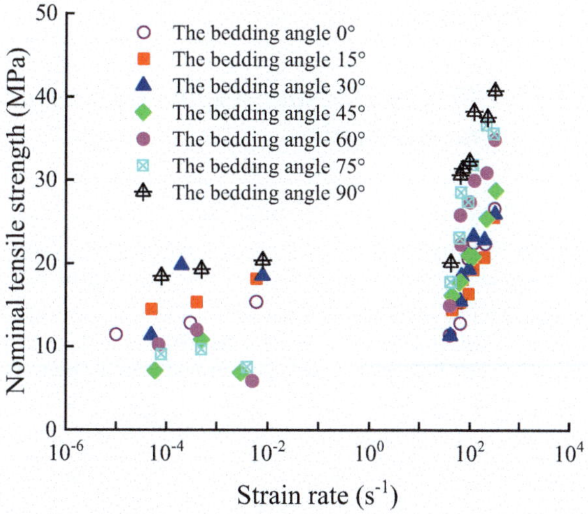

Fig. 2.4 Relationship between the tensile strength and the strain rate for different bedding angles (Feng et al. 2024)

fracture energy, kinetic energy, rotational energy of the resulting debris, or energy in other forms.

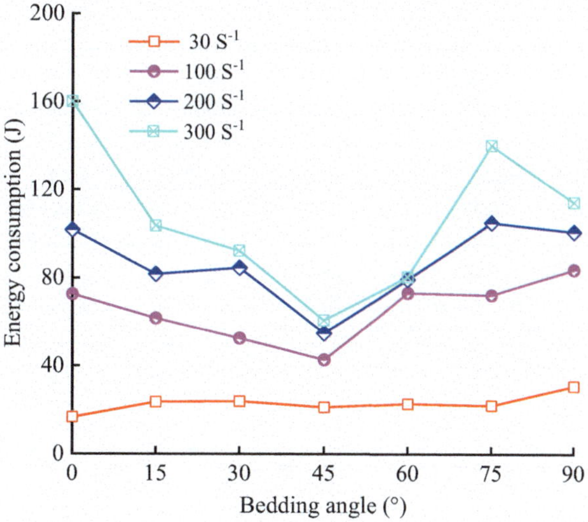

Fig. 2.5 Energy consumption of the specimens under different strain rates (Feng et al. 2024)

2.4 Discussion

2.4.1 Effect of Bedding Angle on Deformation Characteristics

Figure 2.6 illustrate that the bedding angle significantly impacts the failure patterns of shales. In summary, the failure patterns of shales with varying bedding angles can be categorized into four typical types: (a) central fracture, (b) single-layer fracture, (c) Z-type fracture, and (d) parallel-double fractures, as depicted in Fig. 2.10.

The results reveal that as the bedding angle increases, the fracture pattern transitions from a central fracture to a layer-activated mixed mode, with shear failure gradually being replaced by tensile failure. For samples at 0° or 90°, the central fracture mode remains largely unaffected by the bedding plane angle, with a primary crack forming in the loading direction. Between 15° and 75°, the anisotropy of the model specimen becomes increasingly apparent as the bedding angle rises. The 90° specimen exhibits a typical central fracture II that is unaffected by the inclination of the bedding interface. To understand the physical mechanisms behind these four typical fracture types, cross-section fracture morphology analysis was performed using SEM.

By examining micrographs of representative rock fragments at different magnifications, we analyzed the rock fracture characteristics on a microscale. Two distinct categories, cleavage surface and fractured surface (Zhang and Zhao 2013a), were clearly observable. Cleavage fracture is a type of transgranular fracture that predominantly occurs in extensional and tension-shear fractures (Li et al. 2021). The 0° specimen demonstrates a typical central fracture I, where the loading direction aligns with the bedding direction, resulting in cleavage fracture along a crystallographic plane. Additionally, the fracture surface can be seen in the fracture morphologies depicted in Fig. 2.7. The fractured surface of the shale corresponds to the fragile failure behavior of the clay mineral, as shown in Fig. 2.7.

Furthermore, a full-field strain analysis was conducted using the Digital Image Correlation (DIC) technique to investigate the variation in strain distribution on the surfaces of shale specimens with different bedding angles. Figure 2.8 displays the

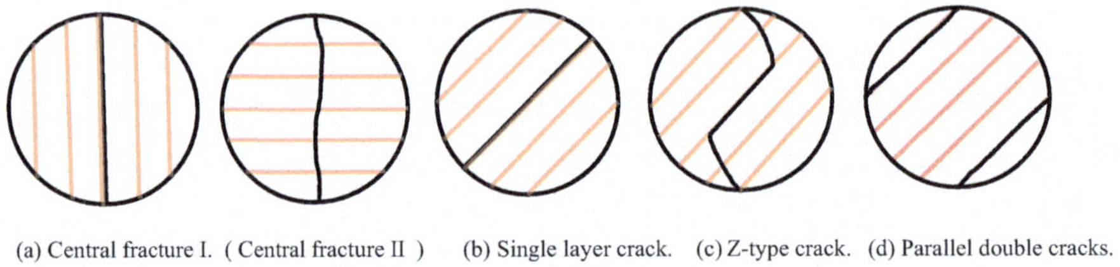

Fig. 2.6 Classification of fracture types: **a** central fracture I; **b** central fracture II; **c** single layer crack; **d** Z-type crack; **e** parallel double crack (Feng et al. 2024)

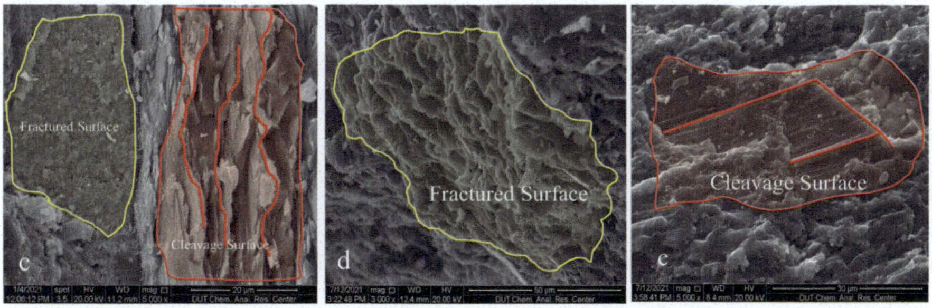

Fig. 2.7 Two typical fracture modes of SEM micrographs fracture. Modified after Feng et al. (2024)

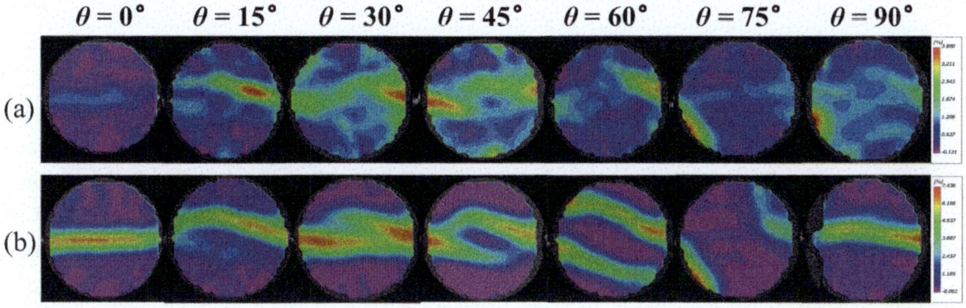

Fig. 2.8 Strain fields of the specimens with different bedding angles. Modified after Feng et al. (2024)

lateral strain contours of the disk surfaces, where positive and negative values represent tensile and compressive strains, respectively. The high tensile strain, indicated by red color, signifies the potential formation of tensile cracks. The strain contours highlight strain localization and the initiation of cracks.

Initially, the deformation field is distributed uniformly, suggesting that the stress inside the specimen gradually achieves a stress equilibrium state due to the multiple reflections of the incident wave within the specimen (Li et al. 2020). Figure 2.8a exhibits localized strains concentrated in the areas adjacent to the load point, with local high strains reaching up to 3.8%, indicating the emergence of plastic deformations within the specimen. These plastic zones contribute to local stretches near the load point, resulting in high local positive strains. As shown in Fig. 2.8b, local strain accumulates rapidly in the area near the impacted side of the sample. Subsequently, the highly concentrated strains continue to propagate across the entire observation area, ultimately reaching a total concentrated strain of 7.4%.

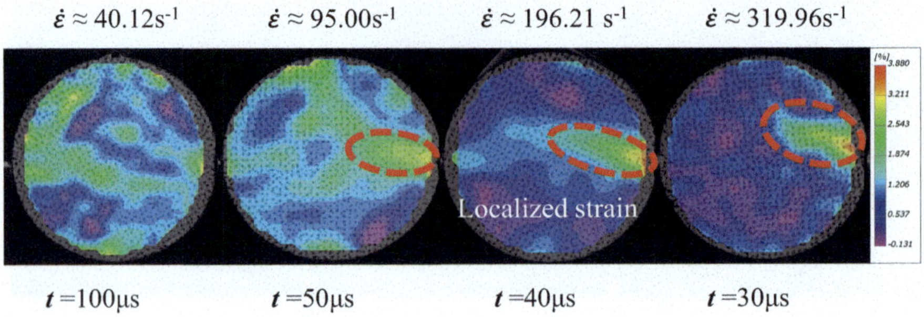

Fig. 2.9 Strain fields of the specimens ($\theta = 30°$) under increasing strain rates. Modified after Feng et al. (2024)

2.4.2 *Effect of Strain Rate on Deformation Features*

To investigate the fracture characteristics of shales subjected to varying strain rate conditions, samples with a bedding angle of θ = 30° were chosen for SEM testing on rock fractures. The results of high-speed photography and DIC were utilized to analyze the correlation between the strain rate and rock failure patterns. The fracture modes observed in the 30° samples exhibited mixed features, including sliding along the bedding interface and shearing across the bedding interface. Notably, the primary cracking direction was parallel to the impact axial, with the axial fracture dividing the disc into two major blocks. The degree of damage in the two shear zones near the load points was closely associated with the load rate. As shown in Fig. 2.9, the magnitude of local strains gradually increased during the initial loading state, and strain concentration could be accelerated with an increasing strain rate. At higher strain rates, stress concentrations at the two ends of the specimen intensified, and the fracture speed of the sample increased more rapidly. Consequently, the crack propagation path became shorter, and the number of cracks increased.

2.4.3 *Effect of Strain Rate on Strength*

The dynamic tensile failure is a notable characteristic of anisotropic rock. To establish a strength criterion capable of predicting the dynamic bearing capacity of transversely isotropic shales, we investigated three prominent strength criteria: the Hobbs-Barron criterion (Hobbs 1967), the Nova-Zaninetti criterion (Nova and Zaninetti 1990; Jyh Jong et al. 1997), and the Single Plane of Weakness (SPW) criterion (Lee and Pietruszczak 2008, 2015). To assess the effectiveness of these criteria in predicting the nominal tensile strength of layered shales, we utilized experimental data from three sets of tests, as detailed in Table 2.1.

A comparison between the experimental values and the predicted results based on the different criteria is presented in Fig. 2.10. Additionally, Table 2.1 includes the

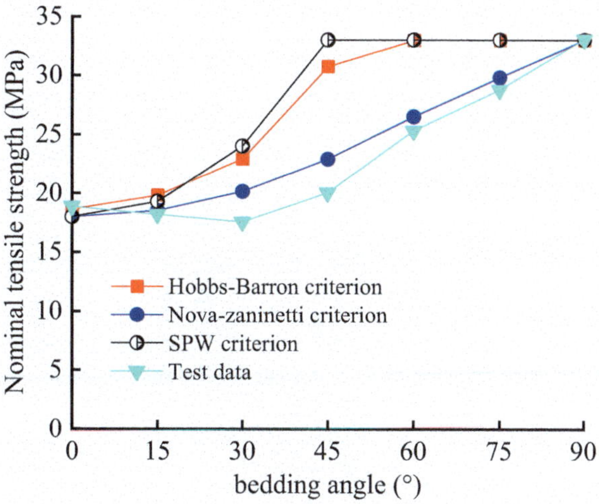

Fig. 2.10 Comparison of the nominal tensile strengths determined by three classical predicted criteria (Feng et al. 2024)

Absolute Relative Error (ARE) for each criterion. The formula for calculating ARE is provided below:

$$ARE = \frac{|T_P(\theta) - T_E(\theta)|}{T_E(\theta)} \times 100\% \quad (2.11)$$

where T_P (θ) is the predicted tensile strength by the various criteria, and T_E (θ) is the mean experimental tensile strength.

Upon examining Table 2.1, it is evident that the predictions of the Nova-Zaninetti criterion align more closely with the experimental data compared to those of the Hobbs-Barron criterion and the SPW criterion. Therefore, in this study, the Nova-Zaninetti criterion has been selected for further refinement to more accurately represent the tensile strength of anisotropic layered shales under dynamic Brazilian splitting. The strength criterion provided by Nova and Zaninetti (1990) can be expressed as follows:

$$T(\theta) = \frac{T_b T_m}{T_b \sin^2(\theta) + T_m \cos^2(\theta)} \quad (2.12)$$

where $\boldsymbol{\theta}$ represents the angle of the applied force direction inclining to the plane of transverse isotropy; $\boldsymbol{T_b}$ represents the tensile strength parallel to the beddings; $\boldsymbol{T_m}$ represents the tensile strength of the intact rock matrix.

Indeed, rock strengths are not static physical constants, as evidenced by previous studies (Qian et al. 2009). These studies have demonstrated that the tensile strength and fracture toughness of rocks increase as the loading rate escalates. Conversely, the static fracture toughness of rocks remains relatively constant (Zhang et al. 2000). Notably, the response of rocks undergoes a significant transformation when the strain rate surpasses a critical threshold (Zhang 2016; Yao et al. 2015). In the current study, a

Table 2.1 Comparison of test results and calculated strengths with different tensile failure criteria (Feng et al. 2024)

Method	Type	$\theta = 0°$	$\theta = 15°$	$\theta = 30°$	$\theta = 45°$	$\theta = 60°$	$\theta = 75°$	$\theta = 90°$	ARE
Test results	No. 1	11.47 MPa	14.55 MPa	11.36 MPa	16.24 MPa	15.00 MPa	17.83 MPa	20.15 MPa	/
	No. 2	12.87 MPa	15.41 MPa	19.83 MPa	17.80 MPa	22.37 MPa	23.17 MPa	30.68 MPa	/
	No. 3	15.43 MPa	18.20 MPa	18.55 MPa	21.08 MPa	27.44 MPa	28.57 MPa	31.48 MPa	/
Test mean	/	18.87 MPa	18.18 MPa	17.55 MPa	20.03 MPa	25.29 MPa	28.76 MPa	33.06 MPa	/
Hobbs-Barron criterion	Strength	18.87 MPa	19.84 MPa	22.95 MPa	30.76 MPa	33.06 MPa	33.06 MPa	33.06 MPa	/
	Error	0.00%	9.14%	30.74%	53.59%	30.68%	14.92%	0.00%	19.80%
Nova-Zaninetti criterion	Strength	18.87 MPa	18.56 MPa	20.30 MPa	16.94 MPa	27.31 MPa	31.00 MPa	33.06 MPa	/
	Error	0.00%	1.92%	14.83%	14.36%	4.81%	3.67%	0.00%	5.66%
SPW criterion	Strength	18.87 MPa	19.29 MPa	24.16 MPa	33.06 MPa	33.06 MPa	33.06 MPa	33.06 MPa	/
	Error	0.00%	6.10%	36.72%	65.07%	30.68%	14.92%	0.00%	21.82%

novel strength criterion that accounts for the strain rate effect has been proposed. This criterion was developed by analyzing Brazilian splitting test data and establishing a dynamic coordinate system.

To quantify the variation in dynamic strength across different strain rates, the Dynamic Increase Factor (DIF) is introduced as a key metric. The DIF is defined as the ratio of dynamic strength to quasistatic strength. To characterize the relative strain rate, a dimensionless quantity $\dot{\varepsilon}/\dot{\varepsilon}_0$ is constructed. This ratio is expressed as a logarithmic function of the strain rate, as follows:

$$DIF = \frac{\sigma_d}{\sigma_0} = K_a \ln\left(m\frac{\dot{\varepsilon}}{\dot{\varepsilon}_0} + 1\right) + b \tag{2.13}$$

$$y = a - bc^x \tag{2.14}$$

When organizing the test data using dimensionless quantities, it becomes apparent that the Dynamic Increase Factor (DIF) in the semi-logarithmic coordinate system correlates with the strain rate. Here, σ_0 denotes the average quasi-static tensile strength when $\theta = 0°$, while σd represents the dynamic tensile strength of the specimen at various strain rates. The symbol $\dot{\varepsilon}$ stands for the strain rate (s^{-1}). In this study, we have set $\dot{\varepsilon}_0 = 3 \times 10^{-6}\ s^{-1}$ for reference. Through fitting, we have determined the coefficients m and b to be 1×10^{-7} and 0.27, respectively.

By fitting the experimental data, we can derive the relationship between the stress intensity factor Ka and the variation in bedding angle as follows:

$$K_a = 2.71 + 0.81 \sin \frac{\pi \times (\theta - \theta_0)}{167} \tag{2.15}$$

where the critical angle constant $\boldsymbol{\theta_0}$ is 92°.

Therefore, a unified equation can be proposed to describe the dynamic strength characteristics of the shales subject to a given strain rate range as follows:

$$DIF = (2.71 + 0.81 \sin \pi(\theta - \theta_0)/167) \times \ln\left(\frac{\dot{\varepsilon}}{\dot{\varepsilon}_0} + 1\right) + 0.27 \tag{2.16}$$

Figure 2.11 shows the DIF curve changing with the strain rate for the shale samples with different bedding directions.

The tensile strength characteristics of anisotropic layered rocks have been discussed by Nova and Zaninetti (1990). By substituting Eq. (2.16) into Eq. (2.12), the dynamic fracture pressure model under the Nova-Zaninetti criterion can be expressed as follows:

$$T(\theta) = \frac{T_b T_m \left[K_a \ln\left(m\frac{\dot{\varepsilon}}{\dot{\varepsilon}_0} + 1\right) + b\right]}{T_b \sin^2\theta + T_m \cos^2\theta} \tag{2.17}$$

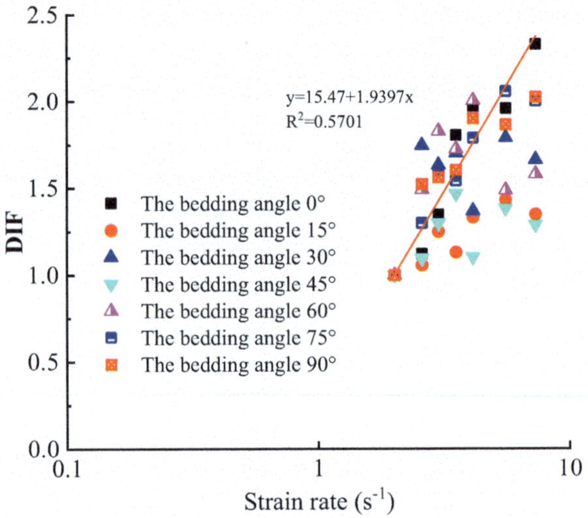

Fig. 2.11 Variation of DIF with the strain rate for the shale samples with different bedding directions (Feng et al. 2024)

To validate the accuracy of the established dynamic strength criterion, we compared the model with experimental data. To quantify the fit of the dynamic strength criteria to shale samples with different bedding angles within a specific strain rate range, we used both the standard deviation and the correlation coefficient R^2 as evaluation metrics.

The results indicate that the strength criteria developed in this paper align well with the test data. When compared to the findings of other researchers, it is evident that the trend in tensile strength variation at different angles with increasing strain rate aligns with the general pattern of dynamic strength changes in shales (Zhang 2016; Zhang and Zhao 2013b). Based on the above analysis, it is clear that both the dynamic tensile strength and the anisotropy of failure patterns increase as the strain rate rises. The predictive model demonstrates high accuracy for samples where anisotropy is the dominant fracture mode, suggesting that the proposed model has a solid physical foundation and is applicable across a wide range of strain rates (30–300 s^{-1}). The fitting formula takes into account the dynamic strength test law, offers a fresh perspective on the rate effect, and can be effectively used to calculate the dynamic strength of layered shales.

2.5 Conclusion

The deformation and failure behaviors of anisotropic shale specimens subjected to various impact loads were investigated using Split Hopkinson Pressure Bar (SHPB) testing equipment and high-speed Digital Image Correlation (DIC) technology. Systematic experimental studies were conducted to reveal the failure process, fracture modes, and mechanical mechanisms of shales under dynamic impact at both macroscopic and microscopic scales. The failure patterns and strength characteristics were comprehensively discussed, leading to the following main conclusions:

(1) The crack initiation position in disk specimens is significantly influenced by both the bedding angle and the strain rate. Four classic failure patterns were observed: (a) central fracture I and central fracture II; (b) single-layer crack; (c) Z-type crack; (d) parallel-double cracks.
(2) Cracks primarily form along the loading direction, and the fracture patterns are only slightly affected by the strain rate when $\theta = 0°$ and $\theta = 90°$. The strain rate effect exhibits anisotropic characteristics and becomes more pronounced as the layer orientation changes from 15° to 75°.
(3) At low strain rates, central fractures are more likely to occur, and the orientation of the bedding planes has a minimal influence on the failure pattern. As the strain rate increases, two parallel cracks become apparent along the bedding planes from both ends of the samples. For moderate bedding plane angles, the strain rate effect induces a relatively complex failure pattern. At high strain rates, cracks generally develop along diverse directions inclined to the bedding planes.
(4) At higher strain rates, the energy release becomes quicker and more significant, leading to the emergence of transgranular cracks. Additionally, as the strain rate increases, high stresses concentrate more prominently at both ends of the specimens, resulting in faster fracture speeds and shorter crack propagation paths with an increased number of cracks.
(5) In the semilogarithmic coordinate system, the dynamic strength exhibits a linear relationship with the growth of the dimensionless quantity ($\dot{\varepsilon}/\dot{\varepsilon}_0 + 1$. The proposed dynamic strength formula contains two parameters with clear physical meanings that are straightforward to determine: the critical strain rate and the bedding angle. This formula has proven reliable and effective for predicting the strength of shales under impact loading within a specific range.

References

Aminzadeh, A., Fahimifar, A., Nejati, M.: On Brazilian disk test for mixed-mode I/II fracture toughness experiments of anisotropic rocks. Theoret. Appl. Fract. Mech. **102**, 222–238 (2019)

Chen, B.P., Gong, B., Wang, S.Y., Tang, C.A.: Research on zonal disintegration characteristics and failure mechanisms of deep tunnel in jointed rock mass with strength reduction method. Mathematics **10**(6), 922 (2022)

Cho, J.W., Kim, H., Jeon, S., Min, K.B.: Deformation and strength anisotropy of Asan gneiss, Boryeong shale, and Yeoncheon schist. Int. J. Rock Mech. Min. Sci. **50**, 158–169 (2012)

Chong, K.P., Boresi, A.P.: Strain rate dependent mechanical properties of new albany reference shale. Int. J. Rock Mech. Min. Sci. Geomech. Abs. **27**, 199–205 (1990)

Debecker, B., Vervoort, A.: Experimental observation of fracture patterns in layered slate. Int. J. Fract. **159**, 51–62 (2009)

Dong, G., Chen, P., Yuan, H., Lu, Y.: The experimental investigation of longmaxi shale dynamic parameters under water-based mud soaking. Geofluids **2019**, 2128373 (2019)

Feng, X.H., Gong, B., Cheng, X.F., Zhang, H.H., Tang, C.A.: Anisotropy and microcrack-induced failure precursor of shales under dynamic splitting. Geomat. Nat. Haz. Risk **13**(1), 2864–2889 (2022a)

Feng, X.H., Gong, B., Tang, C.A., Zhao, T.: Study on the non-linear deformation and failure characteristics of EPS concrete based on CT-scanned structure modelling and cloud computing. Eng. Fract. Mech. **261**, 108214 (2022b)

Feng, X.H., Gong, B., Liang, Z.Z., Wang, S.Y., Tang, C.A., Li, H., Ma, T.H.: Study of the dynamic failure characteristics of anisotropic shales under impact Brazilian splitting. Rock Mech. Rock Eng. **57**, 2213–2230 (2024)

Gong, B., Wang, S.Y., Sloan, S.W., Shen, D.C., Tang, C.A.: Modelling rock failure with a novel continuous to discontinuous method. Rock Mech. Rock Eng. **52**(9), 3183–3195 (2019a)

Gong, B., Tang, C.A., Wang, S.Y., Bai, H.M., Li, Y.C.: Simulation of the nonlinear mechanical behaviors of jointed rock masses based on the improved discontinuous deformation and displacement method. Int. J. Rock Mech. Min. Sci. **122**, 104076 (2019b)

Gong, B., Wang, Y.Y., Zhao, T., Tang, C.A., Yang, X.Y., Chen, T.T.: AE energy evolution during CJB fracture affected by rock heterogeneity and column irregularity under lateral pressure. Geomat. Nat. Haz. Risk **13**(1), 877–907 (2022)

Gholami, R., Rasouli, V.: Mechanical and elastic properties of transversely isotropic slate. Rock Mech. Rock Eng. **47**, 1763–1773 (2013)

Heng, S., Guo, Y., Yang, C., Daemen, J.J.K., Li, Z.: Experimental and theoretical study of the anisotropic properties of shale. Int. J. Rock Mech. Min. Sci. **74**, 58–68 (2015)

Hobbs, D.W.: Rock tensile strength and its relationship to a number of alternative measures of rock strength. Int. J. Rock Mech. Min. Sci. Geomech. Abs. **4**, 115–127 (1967)

Jaeger, J.C.: Shear failure of anisotropic rocks. Geol. Mag. **97**, 65–72 (2009)

Jing, L.: A review of techniques, advances and outstanding issues in numerical modelling for rock mechanics and rock engineering. Int. J. Rock Mech. Min. Sci. **40**, 283–353 (2003)

Jyh Jong, L., Yang, M.T., Hsieh, H.Y.: Direct tensile behavior of a transversely isotropic rock. Int. J. Rock Mech. Min. Sci. **34**, 837–849 (1997)

Pan, R., Zhang, G., Li, S., Zheng, X., Xu, C., Fan, Z.: Influence of the fracture process zone on fracture propagation mode in layered rocks. J. Petrol. Sci. Eng. **202**, 108524 (2021)

Perras, M.A., Diederichs, M.S.: A Review of the tensile strength of rock: concepts and testing. Geotech. Geol. Eng. **32**, 525–546 (2014)

Pufahl, P.K., Anderson, S.L., Hiatt, E.E.: Dynamic sedimentation of Paleoproterozoic continental margin iron formation, Labrador Trough, Canada: Paleoenvironments and sequence stratigraphy. Sediment. Geol. **309**, 48–65 (2014)

Qian, Q., Qi, C., Wang, M.: Dynamic strength of rocks and physical nature of rock strength. J. Rock Mech. Geotech. **1**, 1–10 (2009)

Raef, A.E., Kamari, A., Totten, M., Harris, D., Janssen, K., Lambert, M.: The dynamic elastic and mineralogical brittleness of woodford shale of the Anadarko Basin: ultrasonic P-wave and S-wave velocities, XRD-mineralogy and predictive models. J. Petrol. Sci. Eng. **169**, 33–43 (2018)

Ramamurthy, T.: Strength, modulus responses of anisotropic rocks. In: Hudson, J.A. (ed.) Compressive Rock Engineering, vol. 1, pp. 313–329. Pergamon, Oxford (1993)

Sharafisafa, M., Aliabadian, Z., Shen, L.: Crack initiation and failure development in bimrocks using digital image correlation under dynamic load. Theoret. Appl. Fract. Mech. **109**, 102688 (2020)

Szwedzicki, T.: A hypothesis on modes of failure of rock samples tested in uniaxial compression. Rock Mech. Rock Eng. **40**, 97–104 (2006)

Tavallali, A., Vervoort, A.: Effect of layer orientation on the failure of layered sandstone under Brazilian test conditions. Int. J. Rock Mech. Min. Sci. **47**, 313–322 (2010)

Wang, Y.Y., Gong, B., Zhang, Y.J., Yang, X.Y., Tang, C.A.: Progressive fracture behavior and acoustic emission release of CJBS affected by joint distance ratio. Mathematics **10**(21), 4149 (2022a)

Wang, Y.Y., Gong, B., Tang, C.A., Zhao, T.: Numerical study on size effect and anisotropy of columnar jointed basalts under uniaxial compression. Bull. Eng. Geol. Env. **81**, 41 (2022b)

Wang, Y.Y., Gong, B., Tang, C.A.: Numerical investigation on anisotropy and shape effect of mechanical properties of columnar jointed basalts containing transverse joints. Rock Mech. Rock Eng. **55**, 7191–7222 (2022c)

Lee, Y.K., Pietruszczak, S.: Application of critical plane approach to the prediction of strength anisotropy in transversely isotropic rock masses. Int. J. Rock Mech. Min. Sci. **45**, 513–523 (2008)

Lee, Y.K., Pietruszczak, S.: Tensile failure criterion for transversely isotropic rocks. Int. J. Rock Mech. Min. Sci. **79**, 205–215 (2015)

Lepique, L.B.M.: Empfehlung Nr. 10 des Arbeitskreises 3.3 "Versuchstechnik Fels" der Deutschen Gesellschaft für Geotechnik e.V.: Indirekter Zugversuch an Gesteinsproben—Spaltzugversuch. Bautechnik **85**, 623–627 (2008)

Li, C., Hu, Y., Meng, T., Zhang, C., Gao, R., Jin, P., Hu, Y.: Mode-I fracture toughness and mechanisms of salt-rock gypsum interlayers under real-time high-temperature conditions. Eng. Fract. Mech. **240**, 107357 (2020)

Li, D., Ma, J., Wan, Q., Zhu, Q., Han, Z.: Effect of thermal treatment on the fracture toughness and subcritical crack growth of granite in double-torsion test. Eng. Fract. Mech. **253**, 107903 (2021)

Liang, Z.Z., Xiao, D.K., Li, C.C., Wu, X.K., Gong, B.: Numerical study on strength and failure modes of rock mass with discontinuous joints. Chin. J. Geotech. Eng. **36**(11), 2086–2095 (2014)

Liang, Z.Z., Gong, B., Li, W.: Instability analysis of a deep tunnel under triaxial loads using a three-dimensional numerical method with strength reduction method. Tunn. Undergr. Space Technol. **86**, 51–62 (2019)

Lifshitz, J.M., Leber, H.: Data processing in the split Hopkinson pressure bar tests. Int. J. Impact Eng. **15**, 723–733 (1994)

Liu, X., Dai, F., Zhang, R., Liu, J.: Static and dynamic uniaxial compression tests on coal rock considering the bedding directivity. Environ. Earth Sci. **73**, 5933–5949 (2015)

Ma, T., Peng, N., Zhu, Z., Zhang, Q., Yang, C., Zhao, J.: Brazilian tensile strength of anisotropic rocks: review and new insights. Energies **11**(2), 304 (2018)

Mahanta, B., Singh, T.N., Ranjith, P.G., Vishal, V.: Experimental investigation of the influence of strain rate on strength; failure attributes and mechanism of Jhiri shale. J. Nat. Gas Sci. Eng. **58**, 178–188 (2018)

Nath, F., Mokhtari, M.: Optical visualization of strain development and fracture propagation in laminated rocks. J. Petrol. Sci. Eng. **167**, 354–365 (2018)

Niu, C., Zhu, Z., Zhou, L., Li, X., Ying, P., Dong, Y., Deng, S.: Study on the microscopic damage evolution and dynamic fracture properties of sandstone under freeze-thaw cycles. Cold Reg. Sci. Technol. **191**, 103328 (2021)

Nova, R., Zaninetti, A.: An investigation into the tensile behavior of a schistose rock. Int. J. Rock Mech. Min. Sci. Geomech. Abs. **27**, 231–242 (1990)

Qiu, J., Li, D., Li, X.: Dynamic failure of a phyllite with a low degree of metamorphism under impact Brazilian test. Int. J. Rock Mech. Min. Sci. **94**, 10–17 (2017)

Wu, R.J., Li, H.B., Wang, D.P.: Full-field deformation measurements from Brazilian disc tests on anisotropic phyllite under impact loads. Int. J. Impact Eng. **149**, 103790 (2021)

Yang, G., Li, X., Bi, J., Cheng, S.: Dynamic crack initiation toughness of shale under impact loading. Energies **12**, 1636 (2019)

Yang, S.Q., Yin, P.F., Li, B., Yang, D.S.: Behavior of transversely isotropic shale observed in triaxial tests and Brazilian disc tests. Int. J. Rock Mech. Min. Sci. **133**, 104435 (2020)

Yao, Y., Hu, J., Zhou, A., Luo, T., Wang, N.: Unified strength criterion for soils, gravels, rocks, and concretes. Acta Geotech. **10**, 749–759 (2015)

Yu, C.Y., Gong, B., Wu, N., Xu, P.L., Bao, X.K.: Simulation of the fracturing process of inclusions embedded in rock matrix under compression. Appl. Sci. **12**(16), 8041 (2022)

Zhang, Q.B., Zhao, J.: Effect of loading rate on fracture toughness and failure micromechanisms in marble. Eng. Fract. Mech. **102**, 288–309 (2013a)

Zhang, Q.B., Zhao, J.: A review of dynamic experimental techniques and mechanical behaviour of rock materials. Rock Mech. Rock Eng. **47**, 1411–1478 (2013b)

Zhang, Z.X.: Rock fracture and blasting, chapter 6 of environmental effects on rock fracture. In: Theory applications, pp. 135–153 (2016)

Zhang, Z.X., Kou, S.Q., Jiang, L.G., Lindqvist, P.A.: Effects of loading rate on rock fracture: fracture characteristics and energy partitioning. Int. J. Rock Mech. Min. Sci. **37**, 745–762 (2000)

Zhao, J.: Applicability of Mohr-Coulomb and Hoek-Brown strength criteria to the dynamic strength of brittle rock. Int. J. Rock Mech. Min. Sci. **37**, 1115–1121 (2000)

Zhao, P., Xie, L.Z., Fan, Z.C., Deng, L., Liu, J.: Mutual interference of layer plane and natural fracture in the failure behavior of shale and the mechanism investigation. Pet. Sci. **18**, 618–640 (2021)

Zhou, J., Zeng, Y.J., Guo, Y.T., Chang, X., Liu, L.W., Wang, L., Hou, Z.K., Yang, C.H.: Effect of natural filling fracture on the cracking process of shale Brazilian disc containing a central straight notched flaw. J. Petrol. Sci. Eng. **196**, 107993 (2021)

Zhu, Q., Ma, C., Li, X., Li., D.: Effect of filling on failure characteristics of diorite with double rectangular holes under coupled static-dynamic loads. Rock Mech. Rock Eng. **54**(6), 2741–2761 (2021)

Zhu, Q., Li, D., Han, Z., Xiao, P., Li, B.: Failure characteristics of brittle rock containing two rectangular holes under uniaxial compression and coupled static-dynamic loads. Acta Geotech. **17**, 131–152 (2022)

Chapter 3
Anisotropy of Shale Under SHPB Impact Compression

Abstract Understanding the fracture behavior of shale rocks subjected to impact loading is fundamental for the efficient extraction and utilization of geological energy resources. In this research, we utilized the dynamic Split Hopkinson Pressure Bar (SHPB) system, coupled with high-speed camera imagery, to meticulously investigate the progressive failure mechanisms in layered shale. Our study conducted a thorough examination of the dynamic mechanical properties and fracture characteristics of these shales under impact compression conditions. The results indicate that, based on the initiation mechanisms of microcracks and the mechanical factors influencing fracture patterns, the failure modes of shale rocks can be distinctly classified into five types. Moreover, our analysis highlights that the bedding angle (θ) plays a crucial role in determining the failure mode of the shale samples. Specifically: At $\theta = 0°$, the sample experiences splitting failure, with acoustic emissions primarily concentrated at the horizontal bedding planes. At $\theta = 30°$ and 60°, the sample exhibits shear failure, characterized by a combination of shear cracks, mixed shear cracks, and multiple composite tensile-shear cracks. At $\theta = 90°$, the sample displays both tensile and splitting cracks. Furthermore, as the strain rate increases, the fracture pattern of the sample with $\theta = 60°$ shifts predominantly towards slip and tensile-shear mixed cracks. As the peak stress intensity increases, the failure modes of the samples with $\theta = 30°$ and 60° gradually transition from center-oriented failure to shear failure along the bedding planes. Notably, for the sample with $\theta = 90°$, as the intensity of the shock wave increases, secondary splitting cracks emerge within the matrix, suggesting a fracture pattern that transcends the bedding planes. These findings offer valuable insights into the complex failure mechanisms of shale rocks under dynamic loading conditions, which are crucial for advancing the field of geological energy exploration and utilization.

Keywords Shale failure · Bedding inclination · SHPB · Dynamic anisotropy · Strain rate effect

© The Author(s), under exclusive license to Springer Nature Singapore Pte Ltd. 2025

X. Feng and B. Gong, *Mechanical Properties of Layered Rock Mass*,
https://doi.org/10.1007/978-981-96-5142-9_3

3.1 Introduction

A wide array of geotechnical engineering endeavors, encompassing engineering blasting, protective engineering, and seismic exploration, necessitates a profound understanding of stress wave propagation and attenuation within stratified rock masses. Additionally, these endeavors require an appreciation of the subsequent deformation and failure dynamics under dynamic loads, as highlighted by Feng et al. (2022), Liu et al. (2023), and Feng et al. (2024). Shale exhibits brittle and semi-brittle characteristics in response to peak strength loads, with these responses varying according to bedding inclinations, as demonstrated by Liang et al. (2021), Gao et al. (2023), and Tuzingila et al. (2024). The structural anisotropy of shale formations not only governs their dynamic mechanical properties but also introduces unique aspects in stress wave propagation and attenuation, which differ significantly from those observed in homogeneous rock masses. This anisotropy adds multiple layers of complexity and intricacy to the analysis, as discussed by Luo et al. (2023a), Khmelinin et al. (2023), Si et al. (2024), and Zhang et al. (2022). The anisotropy of shale, resulting from variations in its structural features, profoundly influences failure mechanisms. It determines the directionality and strength of rock properties, thereby impacting fracture propagation, deformation behavior, and overall stability, as evidenced by Alhemdi and Gu (2022), Ma et al. (2023), and Liu et al. (2024). Anisotropic rocks exhibit distinct mechanical responses along different axes, leading to varied failure patterns under stress conditions. These variations have significant implications for engineering projects, such as tunneling, mining, and slope stability assessments. Consequently, tailored approaches are required to mitigate the risks associated with anisotropic rocks, as emphasized by Leroy et al. (2015), Nguyen et al. (2024), and Wang et al. (2024).

In recent years, there has been a substantial increase in both experimental and theoretical research into the anisotropic mechanical characteristics of layered shales. Luo et al. (2023b) and Chao et al. (2024) conducted studies using soft and hard laminated rock-like specimens to pinpoint areas of crack initiation, with notable findings from Brazilian splitting tests. Zhang et al. (2023) further categorized the failure modes of Brazilian shale discs into three distinct types, emphasizing a tendency for tensile failure along the bedding plane. Luo et al. (2022) also adapted a constitutive model to predict shale strength under varying confining pressures more accurately. Kolawole and Oppong (2023) provided valuable micromechanical insights, shedding light on both micro- and macro-mechanical properties of shales. Zhang et al. (2024) examined the significance of shale heterogeneity, identifying it as a key factor influencing critical pressure magnitudes, as evidenced by hydraulic fracturing tests on shales with differing bedding inclinations. Zhao et al. (2023) uncovered distinct micromechanisms governing shale ductility under varying conditions of high confining pressure and temperature through meticulous triaxial compression experiments on bedding shale specimens.

However, the dynamic fracturing process, including crack initiation and development, remains challenging to fully comprehend, and its impact on the impact resistance of layered shales remains unclear. To complement laboratory experiments, numerical simulation has emerged as a powerful tool for gaining deeper insights into the complex mechanisms governing the failure of anisotropic and heterogeneous materials within layered rock formations. Zhou et al. (2024) developed a numerical three-dimensional Split Hopkinson Pressure Bar (SHPB) test, utilizing a combination of particle flow code and fast Lagrangian analysis of continuum methods to replicate the laboratory behavior of shale specimens accurately. Wu et al. (2023) observed a nuanced crack evolution pattern characterized by a concave tendency, leading to a shift in failure mode from matrix-dominated to bedding-dominated, using the discrete element method in bedding sandstone models with two pre-existing flaws. Zhai et al. (2022) analyzed the progressive failure dynamics and acoustic emission signatures of shales using the bonded-particle model integrated with the moment tensor algorithm. Wang et al. (2023a, b, c) introduced an innovative three-dimensional rough discrete fracture network (RDFN3D) model, providing a comprehensive framework to consolidate modeling approaches for complex structural planes, enhancing the fidelity of numerical simulations in capturing real-world behaviors. However, traditional numerical technologies have limitations in characterizing the continuous initiation, propagation, and coalescence of cracks induced by the progressive stress process involving stress buildup, stress release, and stress redistribution.

Given the intricate mechanical dynamics embedded within layered rock masses and the substantial benefits offered by numerical simulation techniques, this study utilizes the SHPB system test to deeply explore the impact of bedding inclinations and shock waves on the initiation and propagation of cracks within shale formations. Additionally, it strives to elucidate the failure process and the underlying mechanisms of bedded shales when subjected to dynamic impacts. To uncover the subtle mechanical intricacies that govern shale failure characteristics, the study integrates the Rock Failure Process Analysis (RFPA) method with digital image processing techniques. This comprehensive approach enables a meticulous recreation of the entire lifecycle of crack initiation, propagation, and penetration within the underlying rock mass model, all under dynamic impact loads. Through systematic analysis, the study seeks to uncover the fundamental principles and laws governing these processes, thereby providing essential data and a robust theoretical framework to comprehend the large-scale engineering stability of layered rock masses.

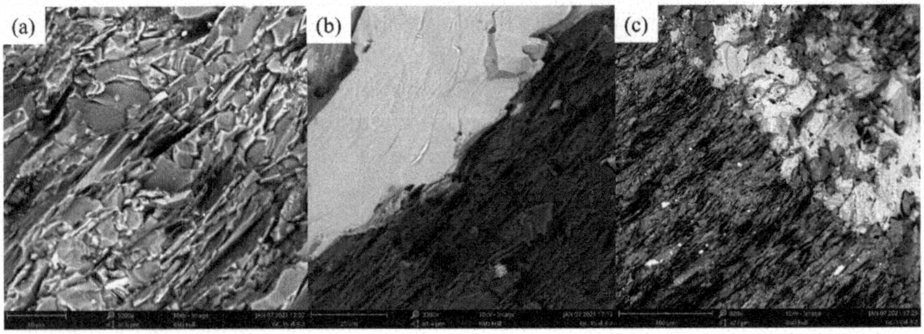

Fig. 3.1 Shale micrograph obtained by the scanning electron microscope (QUANTA450) (Feng et al. 2025)

3.2 Experimental Setup and Process

3.2.1 Sample Preparation

The specimens used in this study were obtained from the Lu Mountain Mine situated in Jiujiang City, Jiangxi Province, China. The stratified rock reservoir in this region boasts well-defined layering and prominent structural stratifications. Figure 3.1a provides a close-up view of the clay minerals, revealing discernible primary defects, particularly microcracks, which are indicative of the inherent heterogeneity of the rock. Figure 3.1b showcases the formation of weak planes arising from the parallel proliferation of microcracks among plate-like minerals, creating vulnerable surfaces within the specimens. On the other hand, Fig. 3.1c illustrates clastic minerals filling the gaps between clay minerals, aligning themselves parallel to the bedding planes. Essentially, the shale specimens selected for this study exhibit a fissile structure at the microscopic scale and a clearly defined layering structure at the macroscopic level, typical of layered rock formations.

To obtain specimens with different bedding dip angles, we first defined the bedding dip angle θ as the angle between the bedding plane direction and the horizontal baseline. Subsequently, using a drill bit with a diameter of 50 mm, we cored the specimens at angles of 0°, 30°, 60°, and 90°, as shown in Fig. 3.2. In accordance with the standards established by the International Society for Rock Mechanics and Engineering (ISRM), the specimens were then processed into cylindrical shapes, with a diameter of 48 mm and a height of 50 mm, as illustrated in Fig. 3.3.

3.2.2 Testing Apparatus and Methodology

The SHPB test system employed in this study, as detailed in previous works by Zhu et al. (2022), Li et al. (2021), and Cai et al. (2021), encompasses three essential

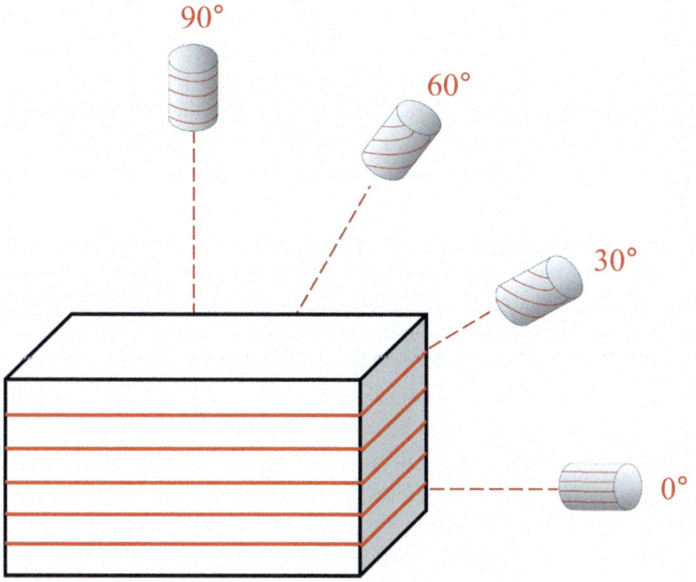

Fig. 3.2 Shale coring schematic (Feng et al. 2025)

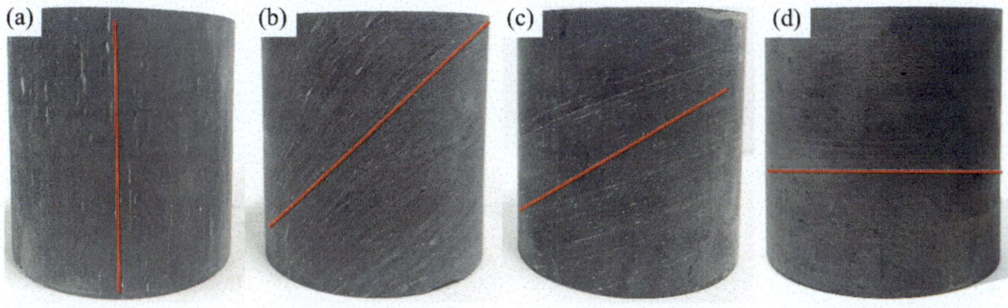

Fig. 3.3 Representative shale specimens with **a** $\theta = 0°$, **b** $\theta = 30°$, **c** $\theta = 60°$ and **d** $\theta = 90°$ (Feng et al. 2025)

components: the dynamic impact system, the stress propagation system, and the signal and data processing system. The experimental setup features an incident bar, a transmission bar, an absorption bar, and a striker, all crafted from silicon-manganese steel with a diameter of 50 mm, a density of 7850 kg/m3, and a longitudinal wave velocity of 5201 m/s. Specifically, the incident bar spans 3.0 m, while the absorption bar and transmission bar measure 1.8 m and 0.5 m in length, respectively.

The dynamic test system incorporates state-of-the-art instrumentation, including a super-dynamic strain gauge (CS-ID), an oscilloscope (DL-750), and a strain gauge model (BX120-2AA), which collectively ensure precise measurement and analysis of stress and strain during the impact process. To visually capture the dynamic failure process of the specimens, a digital image real-time monitoring system is utilized, comprising a high-speed camera and a high-intensity light source. This system captures images at intervals of 10 μs, providing valuable insights into the evolution of cracks and deformation patterns during the experiment.

Due to the disparity in wave impedance between the pressure rod and the tested specimen, the incident stress wave undergoes reflection and transmission at their interface. Upon impact, a portion of the stress wave reflects back into the incident rod, while the remaining energy traverses the specimen and into the transmission rod, completing the loading sequence. Strain gauges affixed to the midpoints of both the incident and transmission rods facilitate the measurement of stress wave propagation components. Specifically, the strain gauge on the incident rod captures the incident stress wave signal $\varepsilon_i(t)$ and the reflected stress wave signal $\varepsilon_r(t)$, while the strain gauge on the transmission rod records the transmitted stress wave signal $\varepsilon_t(t)$.

By designating the contact point between the incident rod and the specimen as X_I and the corresponding contact point between the transmission rod and the specimen as X_2, we can analyze the propagation of stress waves within the pressure rod. Throughout the experiment, stress waves within the pressure rod consistently exhibit elastic behavior. At the critical points X_I and X_2, both stress and particle velocity are measured to elucidate the dynamic response of the system, as follows:

$$\sigma_1(t) = \sigma_1(t) + \sigma_R(t) = E(\varepsilon_1(t) + \varepsilon_R(t)) \tag{3.1}$$

$$v_1(t) = v_1(t) + v_R(t) = C_0(\varepsilon_1(t) - \varepsilon_R(t)) \tag{3.2}$$

$$\sigma_2(t) = \sigma_T(t) = E\varepsilon_T(t) \tag{3.3}$$

$$v_2(t) = v_T(t) = C_0\varepsilon_T(t) \tag{3.4}$$

where $\sigma_1(t)$ and $v_1(t)$ represent the stress and particle velocity at Point X_1, respectively, and $\sigma_2(t)$ and $v_2(t)$ represent the stress and particle velocity at Point X_2, respectively. E and C_0 are the elastic modulus and stress wave velocity of the pressure rod, respectively.

Referring to the equation provided, the average stress $\sigma(t)$, strain rate $\dot{\varepsilon}(t)$, and strain $\varepsilon(t)$ of the specimen can be derived throughout the loading process as follows:

$$\sigma(t) = \frac{EA_0}{2A_s}(\varepsilon_I(t) + \varepsilon_R(t) + \varepsilon_T(t)) = \frac{EA_0}{A_s}\varepsilon_T(t) \tag{3.5}$$

$$\dot{\varepsilon}(t) = \frac{C_0}{L_s}\varepsilon_I(t) - \varepsilon_R(t) - \varepsilon_T(t) = -\frac{2C_0}{L_s}\varepsilon_R(t) \tag{3.6}$$

$$\varepsilon(t) = \int_0^t \dot{\varepsilon}(t)dt = \frac{2C_0}{L_s}\int_0^t \varepsilon_R(t)dt \tag{3.7}$$

In this context, the variables A_0 and As represent the cross-sectional areas of the loading rod and the sample, respectively, whereas L_s signifies the length of the specimen. Additionally, the notations $\sigma_I(t)$, $\sigma_R(t)$, and $\sigma_T(t)$ are used to denote the incident stress, reflected stress, and transmitted stress at any given time t, respectively.

3.3 Experimental Results

In Fig. 3.4, the stress–strain curves of shale specimens subjected to dynamic compression loads are illustrated, each corresponding to different bedding angles. Notably, the application of shock waves induces robust elastic–plastic behavior within the shale specimens. The critical failure curves for each bedding angle highlight the significant impact of shale strength anisotropy, resulting in disparities in peak stress levels. Despite these variations, the overall trend and pattern of the curves remain strikingly similar across the various bedding angles.

During the dynamic impact compression testing, a high-speed camera was utilized to meticulously capture the fracture process occurring within the specimens. Upon the impact of the stress wave, an instantaneous release of energy reverberates through the stratified rock, frequently accompanied by a distinctive loud noise. This crucial moment marks the rapid initiation and propagation of macroscopic cracks within the rock, swiftly traversing the specimen. As these cracks proliferate, the load-bearing capacity of the specimen decreases rapidly, ultimately leading to the formation of a fracture surface.

Figure 3.5 displays the dynamic crack propagation processes in shale specimens with various bedding angles. These images vividly illustrate the morphological differences in failure modes, emphasizing the pronounced anisotropy inherent in shale

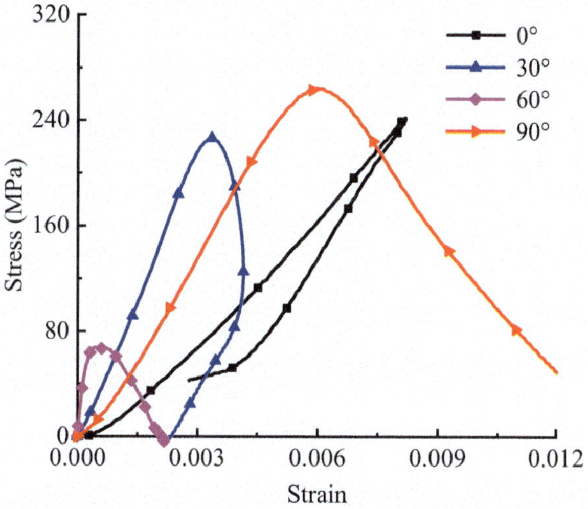

Fig. 3.4 Dynamic stress–strain curves of the shale specimens with different bedding angles (Feng et al. 2025)

materials. In the 0° specimen, an initial horizontal crack emerges along the bedding plane, extending parallel to the axial loading direction. By 140 microseconds (μs), this crack evolves into a macroscopic fissure spanning the entire length of the specimen, indicative of a typical cleavage failure mode. In contrast, the 30° specimen exhibits initial shear cracks along the bedding plane upon the initial impact load, which then extend towards both ends of the specimen. This results in a predominantly shear-dominated tensile-shear mixed failure characteristic. At 90 μs after impact, the 60° specimen displays an initial shear crack along the bedding plane, which further propagates into a macroscopic fissure along the same plane, exhibiting a characteristic shear failure along the bedding plane. The 90° specimen shows an initial crack along the bedding plane. As the shock wave intensifies, secondary splitting cracks emerge in the matrix, indicative of a fracture pattern that crosses the bedding planes.

3.3.1 Failure Pattern and Strain Rate Effect

Throughout the experiment, high-speed photography documented the dynamic failure modes exhibited by the bedded shale specimens. By carefully analyzing the spatial correlation between macroscopic fracture cracks and bedding planes, the failure patterns observed in the shale specimens under impact compression were categorized into five distinct types, as illustrated in Fig. 3.6: (a) splitting failure along the bedding plane (Type I); (b) slip failure along the bedding plane (Type II); (c) mixed shear cracks between layers (Type III); (d) tensile-shear mixed cracks (Type IV); (e) tensile crack along the bedding plane (Type V).

Figure 3.7 depicts the impact fracture features of shale specimens with different bedding angles under various strain rates. Series (a), (b), (c), and (d) correspond to specimens with $\theta = 0°$, 30°, 60°, and 90°, respectively. Meanwhile, Series (1), (2), (3), and (4) represent strain rates of 20 s^{-1}, 80 s^{-1}, respectively. From Fig. 3.10, it is evident that the fracture behaviors of the shale specimens vary significantly based on their bedding angles.

The 0° specimens consistently exhibit a relatively uniform fracture mode, predominantly showcasing a splitting pattern along the horizontal bedding plane, with the fracture surfaces extending along the axial direction. This fracture type is characterized by the penetration of the matrix along the weak bedding plane (Type I). Conversely, the 30° specimens display shear cracks along the bedding planes at both ends (Type II), with the fracture surfaces penetrating the entire shale samples. When the strain rate reaches 100 s^{-1}, the shale specimen exhibits a primary crack along the bedding plane and multiple composite tensile-shear cracks (Type IV), penetrating both the bedding planes and the matrix. Notably, the 30° specimen demonstrates coexisting splitting tensile and shear cracks under all four strain rate conditions.

In contrast, the fracture patterns of the 60° specimens lack consistency. Specifically, when the specimen is subjected to a strain rate of 20 s^{-1}, shear cracking prevails, accompanied by tensile cracks during the testing process (Type II and Type

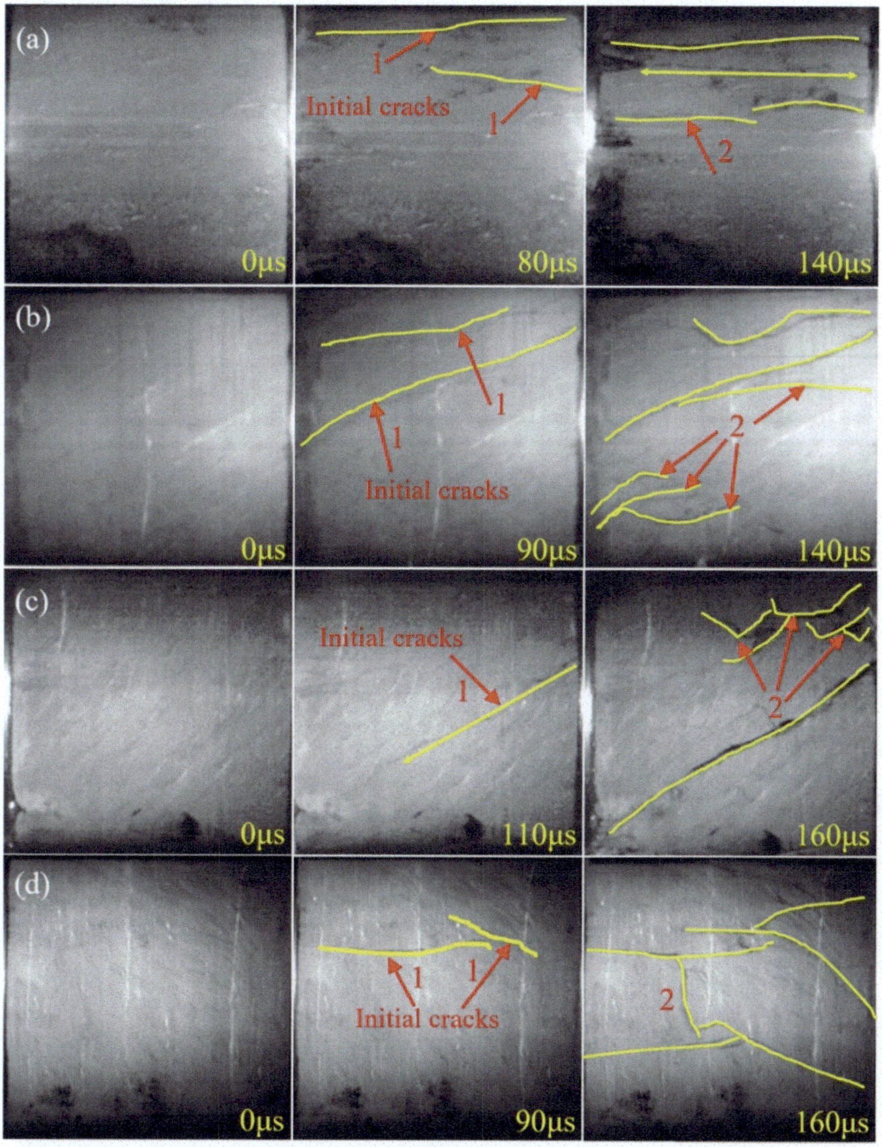

Fig. 3.5 Impact compression testing processes of the specimens with various bedding angles: **a** $\theta = 0°$, **b** $\theta = 30°$, **c** $\theta = 60°$ and **d** $\theta = 90°$. Modified after Feng et al. (2025)

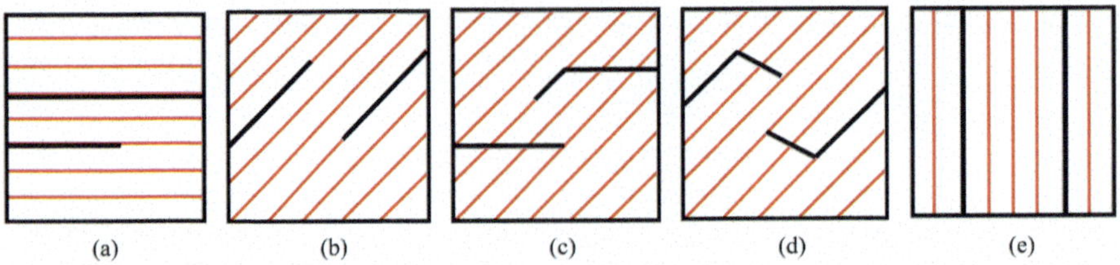

Fig. 3.6 Classification of failure patterns of shales under impact compression: **a** splitting failure along the bedding plane, **b** slip failure along the bedding plane, **c** mixed shear cracks between layers, **d** tensile-shear mixed cracks and **e** tensile crack along the bedding plane. Modified after Feng et al. (2025)

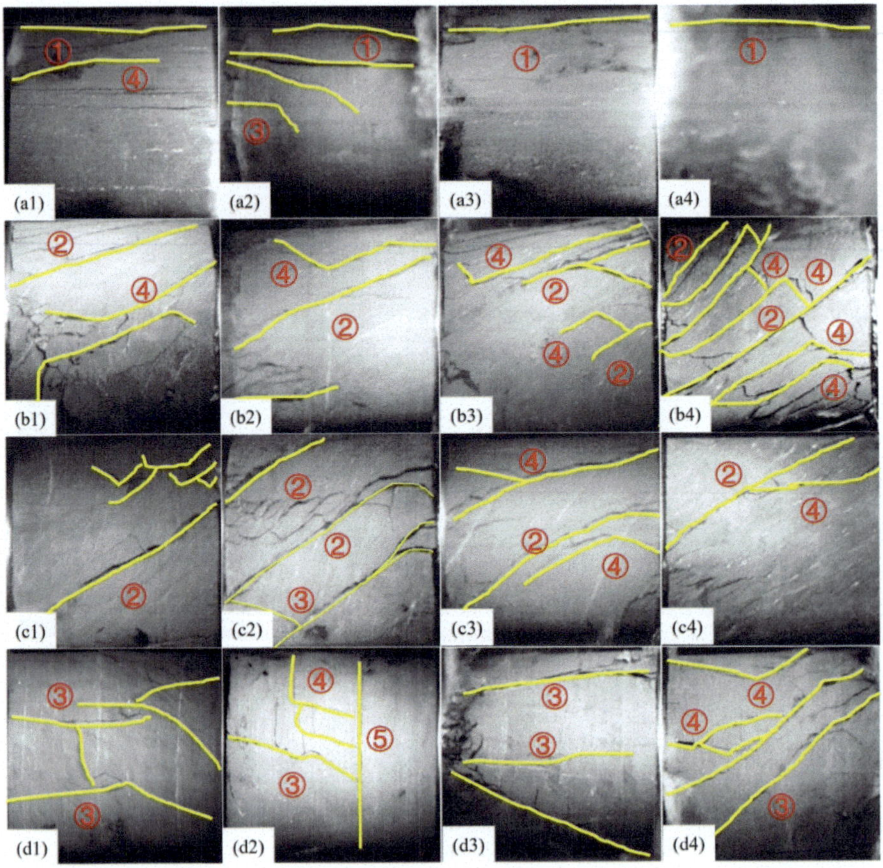

Fig. 3.7 Impact compression failure features of shales. Modified after Feng et al. (2025)

III). With the strain rate continuing to increase, the fracture pattern becomes dominated by a combination of Type II and Type IV cracks. Moreover, both Type II and Type IV cracks coexist across all four strain rates.

When the shale specimen with a 90° bedding angle is subjected to a strain rate of 20 s^{-1}, two Type III tensile cracks penetrate the entire specimen. When the strain rate increases to 50 s^{-1}, Type I tensile cracks are produced along the 90° bedding planes at the transmission end, and one Type V tensile crack can also be observed. As the loaded strain rate continues to grow, the shale exhibits splitting tensile fracture penetrating the matrix and the bedding plane along the loading direction. Finally, when the strain rate equals 100 s^{-1}, large-angle tensile-shear fractures can be measured, penetrating both the matrix and the bedding plane at the loading ends. Simultaneously, the fracture surfaces evolve through multiple bedding planes towards the center of the specimen.

3.4 Numerical Modeling

3.4.1 Determination of Material Segmentation Threshold

In this study, the image threshold segmentation technique was employed to discern the shapes and locations of beddings and the rock matrix, as well as to delineate the finite element mesh (Li et al. 2017; Yu et al. 2018). By applying distinct grey thresholds, various microstructures can be efficiently identified, enabling the determination of different material types. The theoretical foundation for threshold calculation is articulated in Eq. (3.8).

$$f'(x,y)=\begin{cases}1/(\mathrm{n}+1) & f(x,y)\le T_1\\ 2/(\mathrm{n}+1) & T_1\le f(x,y)\le T_2\\ \vdots & \cdots\\ \mathrm{n}/(\mathrm{n}+1) & T_{n-1}\le f(x,y)\le T_n\\ (\mathrm{n}+1)/(\mathrm{n}+1) & f(x,y)>T_n\end{cases} \tag{3.8}$$

Meanwhile, if the grayscale value of a pixel is less than or equal to the threshold T_l, it will be assigned a value of 1/(n + 1). Conversely, if the grayscale value of a pixel exceeds the threshold Tn, its grayscale function will be set to 1. Here, f(x, y) denotes the initial grayscale value of the pixel located at coordinates (x, y), while f(x, y) represents the standardized grayscale value of the same pixel. The grayscale grid within the specified region of the shale model, along with the grayscale values along line AA′, are illustrated in Fig. 3.8.

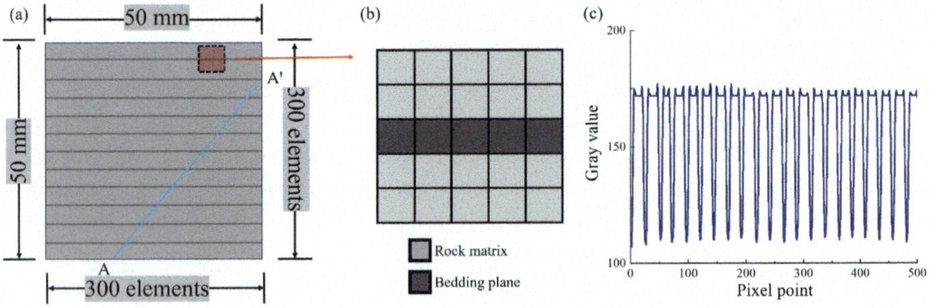

Fig. 3.8 The digital image processing: **a** the captured image, **b** the generated grayscale grid and **c** the grayscale values along the line AA′. Modified after Feng et al. (2025)

3.4.2 Model Configuration

Drawing upon statistical damage mechanics, the RFPA method excels in depicting the incremental cracking behavior of rock masses (Chen et al. 2022; Tang et al. 2020; Wang et al. 2022a). The RFPA has been thoroughly validated by researchers in modeling rock fracture scenarios (Chen et al. 2023; Gong et al. 2022; Wang et al. 2022b; Yu et al. 2022) and was employed for simulations in this study. During the impact load test, a conical bullet head was used, generating a waveform resembling a sine wave. Among various waveforms, the triangle wave closely mirrors the actual wave employed in the experiment, thus being adopted for numerical simulation.

In this section, the model dimensions were set at 50 m × 50 m and divided into 300 × 300 elements, utilizing two-dimensional plane strain constraints for computation, as depicted in Fig. 3.9. A triangle wave was utilized to impact the loading plate over a loading duration of 100 μs. Recognizing the abundance of microjoints and microcracks within the rock mass, which constitutes non-uniform materials, the material non-uniformity was characterized by the Weibull distribution (Gong et al. 2024). The pertinent physical and mechanical parameters are detailed in Table 3.1.

To ensure computational precision, a step-by-step analysis procedure was adopted with specific parameter configurations outlined in the table, allowing the fine cracking process within one time step to be captured across several sub-time steps. To comprehensively examine the impact fracture morphology characteristics of shales, four distinct bedding angles were considered: 0°, 30°, 60°, and 90°. Furthermore, the numerical simulations were conducted in alignment with the aforementioned physical experiments.

3.4.3 Simulation of SHPB Test

Figure 3.10 illustrates the failure processes of models with different bedding angles under dynamic impact compression, presenting both the maximum shear stress

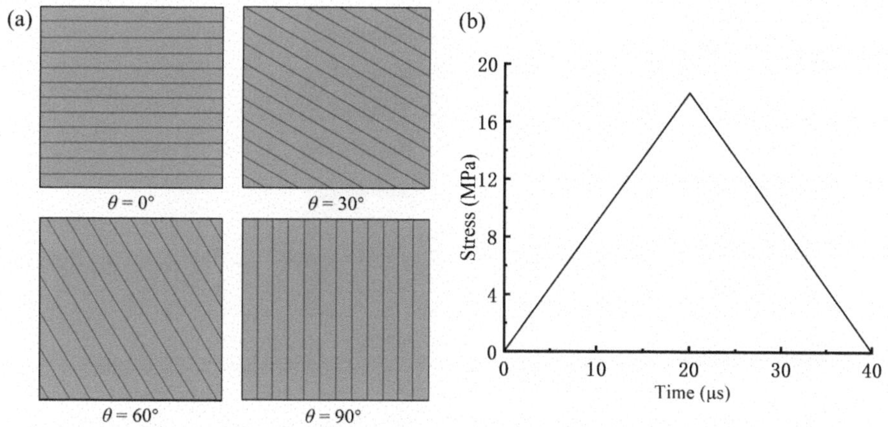

Fig. 3.9 Numerical modeling: **a** the models with various bedding angles. **b** the triangle wave for impact loading. Modified after Feng et al. (2025)

contours and acoustic emission fields. In this figure, the letters *M* and *A* represent the maximum shear stress contours and acoustic emission fields, respectively. The numbers 0, 30, 60, and 90 correspond to the bedding dip angles of 0°, 30°, 60°, and 90°, respectively. Additionally, the centers and radii of the circles indicate the locations and energy magnitudes of acoustic emission events within the acoustic emission fields. The circles colored red and white signify tensile failure mode and compression-shear failure mode, respectively. Regions with denser clusters of circles indicate a higher degree of element damage.

In the 0° bedded model, damaged elements initially appear at the point of impact. As the impact loading progresses, a through crack develops along the bedding plane, extending further along the middle beddings from the initial crack. This numerical simulation aligns with the dynamic impact failure mode observed in the related shale sample in Fig. 3.10, characterized by Type I failure with the failure surface opening along the bedding plane.

During the initial stage of impact loading for the 30° and 60° bedded specimens, micro-cracks appear along the bedding planes near the incident end, accompanied by noticeable damage along the bedding planes. As the impact loading continues, the models experience shear cracks (Type II) and multiple composite tensile-shear cracks (Type IV) along the bedding planes and across the matrix. This simulated result corresponds with the experimental findings of the corresponding shale sample, where damage across the inclined beddings was observed.

In the 60° model, noticeable damage is evident at the edge of the incident end during the initial loading stage. Subsequently, in the later stages of loading, the formation of macroscopic cracks in the rock sample is primarily dominated by shear cracks, accompanied by tensile cracks. The failure mode can be characterized by Type II, Type III, and Type IV cracks, with multiple composite compressive-shear and tensile-shear cracks emerging along the inclined beddings at the center. This

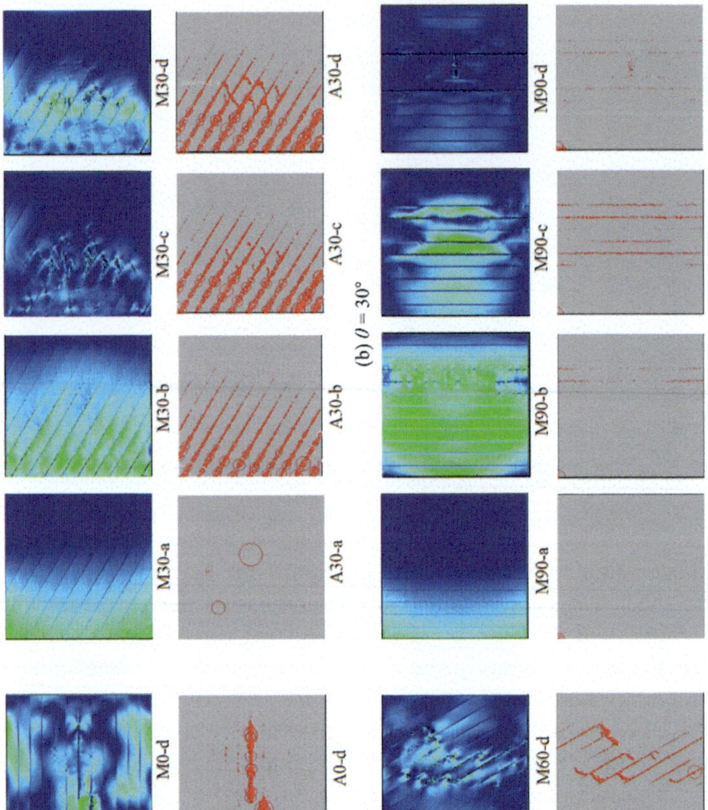

Fig. 3.10 Failure processes of the impact compression models. **a** 0°, **b** 30°, **c** 60° and **d** 90°. Modified after Feng et al. (2025)

Table 3.1 Physical and mechanical parameters for simulation (Feng et al. 2025)

Parameter	Bedding plane	Matrix
Density (kg/m^3)	2500	2500
Young's modulus (GPa)	5	50
Poisson's ratio	0.2	0.2
Homogeneity index	3	8
Unixial compressive strength (MPa)	50	100

outcome closely matches the physical test data, where the central portion of the sample undergoes a mixed fracture mode crossing the bedding planes.

When the bedding angle increases to 90°, tensile fractures occur along the vertical beddings at the middle and right side of the sample due to the tensile wave produced by the superposition of the incident stress wave and reflected stress wave. This phenomenon can also be observed in Fig. 3.10d2.

3.4.4 *Effect of Stress Wave on Rock Fracture*

Both the physical experiment and numerical simulation tests have demonstrated a significant impact of bedding angle and peak strength on the failure characteristics of the samples. Therefore, this section delves deeper into the interplay between peak strength, bedding angle, and the resulting failure morphology of shale samples by employing simulation scenarios at $\theta = 0°$, 30°, 60°, and 90°. To ensure consistency, the model dimensions and mesh configurations used in these scenarios are identical to those employed previously. The loading waveforms corresponding to these scenarios are depicted in Fig. 3.11.

Figure 3.12 showcases the fracture patterns in shale samples exposed to different bedding angles and impact waves. At a bedding angle of 0°, the fractures predominantly appear as Type I cracks, closely mirroring the experimental outcomes depicted in Fig. 3.7. As the bedding angle increases to 30°, Fig. 3.12b3 from the numerical simulation closely resembles the characteristics of the impact test shown in Fig. 3.7b3, with slip cracks along the bedding plane and a dominance of inter-layer mixed shear cracks forming the macroscopic fissures. Similarly, Fig. 3.12b4 aligns with Fig. 3.7b4 from the impact test, where Type II shear cracks emerge at the sample ends, accompanied by multiple Type IV composite tensile-shear damage cracks crossing the bedding plane and penetrating the matrix. When the bedding angle reaches 60°, the fracture modes captured in the numerical simulations reflect the records of the impact tests, highlighting Type II shear cracks, Type III mixed shear cracks, and Type IV tensile-shear mixed cracks as the dominant features, with a distribution and composition similar to those observed in physical experiments. At a 90° bedding angle, all four numerical models produce analogous outcomes characterized by Type V tensile cracks, which resemble the fracture outcome shown in Fig. 3.7d2 from the impact testing.

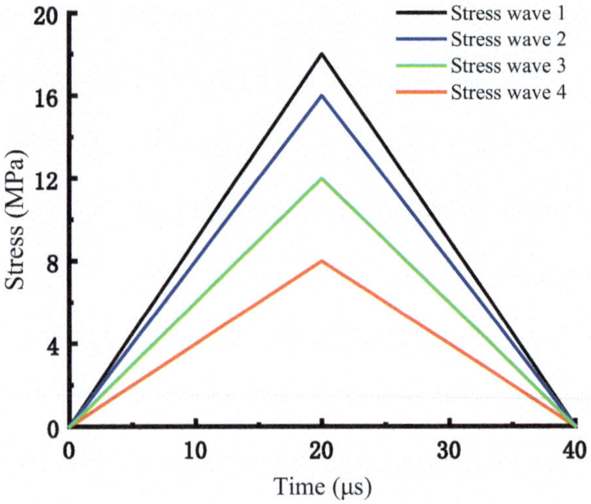

Fig. 3.11 Dynamic impact compressive stress waves. Modified after Feng et al. (2025)

3.4.5 Effect of Bedding Inclination on Rock Fracture

Numerous geotechnical engineering projects, such as blasting and protective construction endeavors, critically rely on an understanding of the dynamics of stress wave propagation, superposition, and attenuation within layered rock formations. As depicted in Fig. 3.13, when the bedding angle θ is 0°, the model undergoes a splitting failure that traverses the bedding planes, with acoustic emissions predominantly concentrated at the horizontal bedding planes located centrally within the sample. The distribution of macroscopic cracks appears relatively straightforward, primarily exhibiting Type I fissures that extend along the weakened bedding planes.

When the bedding angle θ is 30° or 60°, the shale sample primarily experiences shear failure, characterized by the propagation of Type II shear cracks, Type III mixed shear cracks, and multiple Type IV composite tensile-shear cracks along the bedding planes and through the matrix. As the peak stress intensity increases, the failure modes of the shale samples with 30° and 60° bedding angles gradually shift from center-oriented failure to shear failure along the bedding planes, as illustrated in Fig. 3.14a–f. At smaller bedding angles, nearly all cracks within the sample appear as shear fractures along the bedding planes, often accompanied by localized tensile failures. This suggests that crack initiation at bedding planes may be uncommon, with failures primarily originating within the rock matrix. Based on the classification of failure modes in layered rock masses, it is evident that variations in bedding angle and peak strength significantly influence the stability of rock structures. Figure 3.14g and h further elucidate the failure patterns of rock structures, highlighting the substantial impact of changes in bedding plane angle on the overall failure pattern of the rock structure.

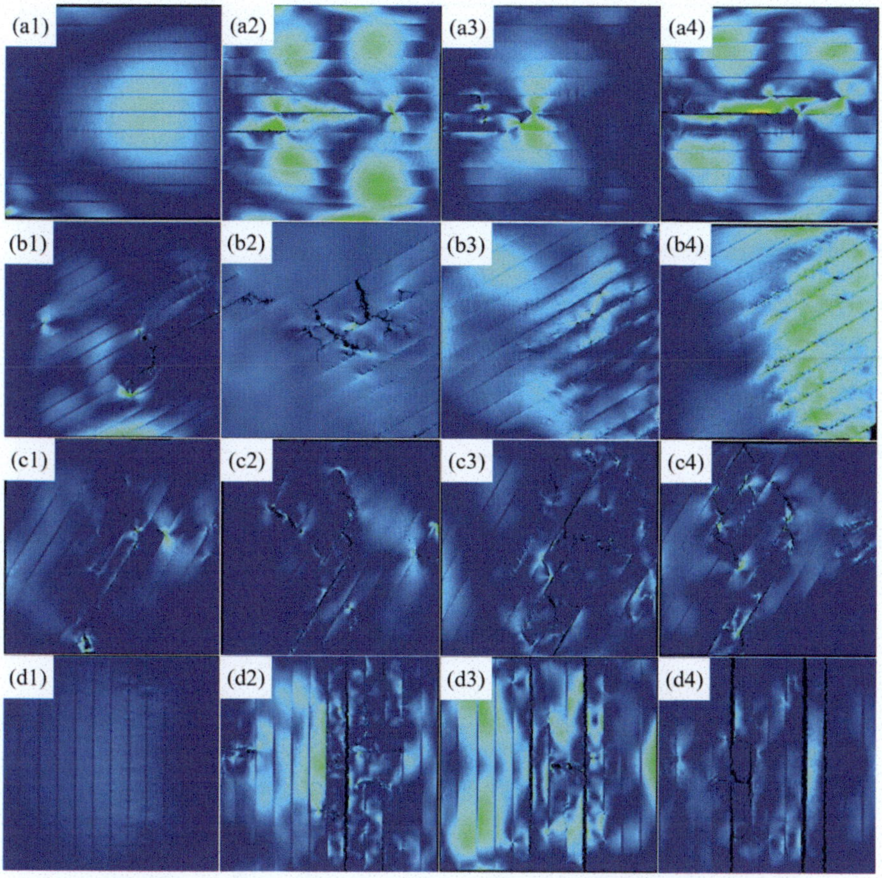

Fig. 3.12 The failure modes represented by the shear stress contours. Modified after Feng et al. (2025)

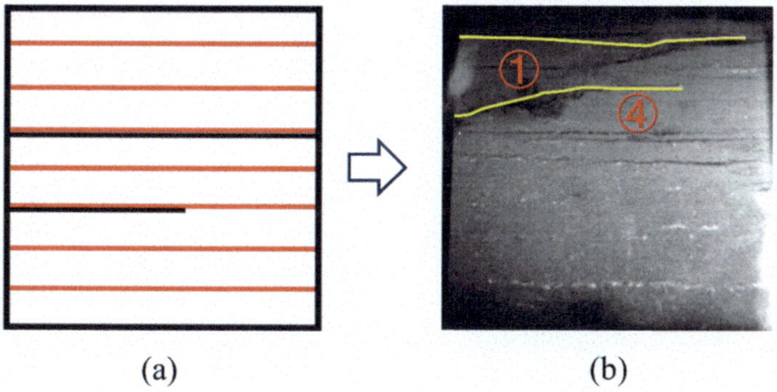

Fig. 3.13 Fracture characteristics of the shale sample with the bedding angle of 0°: **a** schematic Type I cracks and **b** cracks observed in test. Modified after Feng et al. (2025)

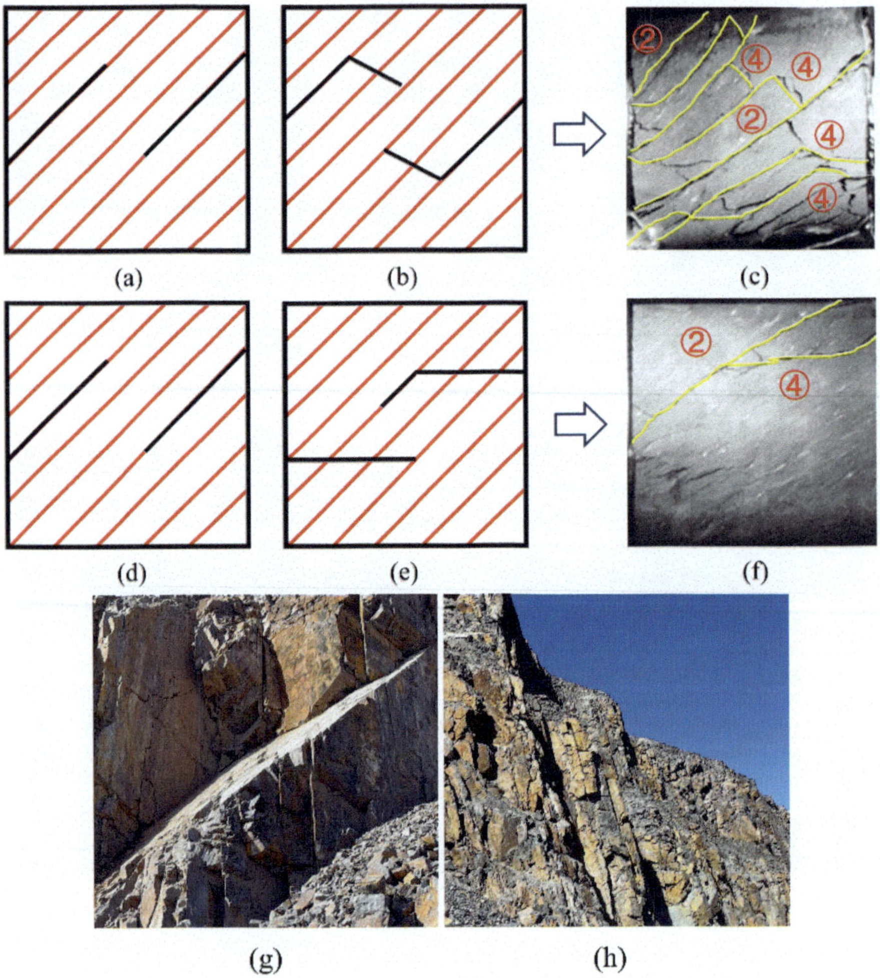

Fig. 3.14 Failure characteristics of the rock mass with inclined structural planes: **a**–**b**, **d**–**e** schematic Type II, Type III and Type IV cracks; **c**, **f**: cracks observed in test; **g**–**h** cracks observed on-site. Modified after Feng et al. (2025)

Figure 3.15 demonstrates that when the bedding angle θ is 90°, the sample displays tensile and splitting cracks that propagate along its bedding planes, ultimately penetrating the entire sample. As the peak stress wave intensity increases, the primary macroscopic cracks within the 90° bedded shale sample remain relatively consistent in their fracture characteristics, exhibiting minimal variation. However, there is a notable increase in the number of smaller cracks forming between the bedding planes.

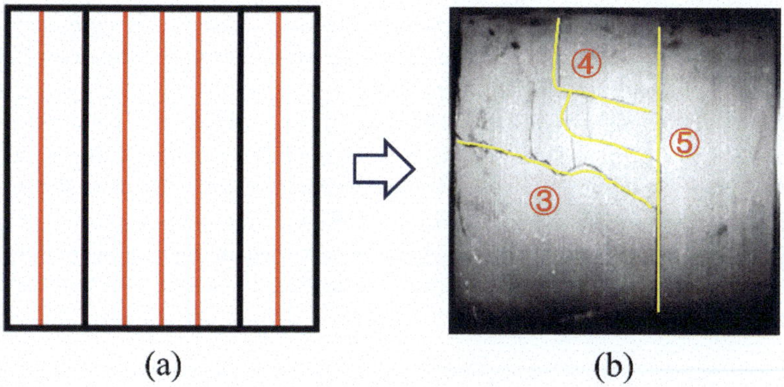

Fig. 3.15 Fracture characteristics of the shale sample with the bedding angle of 90°: **a** schematic Type V cracks and **b** cracks observed in test. Modified after Feng et al. (2025)

3.5 Conclusion

Uncovering the fracture characteristics of bedded shales under dynamic impact loads is crucial for the exploitation and utilization of geological energy. In this study, the progressive failure processes of bedded shale samples were examined using a dynamic SHPB system and monitored by a high-speed camera. Furthermore, the relevant dynamic mechanical properties and fracture features inherent in shale formations under impact compression were thoroughly investigated. The primary conclusions drawn from this investigation are outlined below:

(1) By rigorously analyzing the triggering mechanisms of micro-cracks and the mechanical factors governing fracture morphology, the failure patterns of shale specimens under impact compression can be categorized into five distinct types: (a) splitting failure along the bedding plane; (b) slip failure along the bedding plane; (c) mixed shear cracks between layers; (d) tensile-shear mixed cracks; and (e) tensile crack along the bedding plane.
(2) The dynamic fracture behaviors of shale specimens vary significantly based on their bedding angles. Specifically, when the bedding angle θ is 0°, the model undergoes a splitting failure that traverses the bedding planes, with acoustic emissions predominantly concentrated at the horizontal bedding planes. At $\theta =$ 30° and 60°, the shale sample experiences shear failure characterized by shear cracks, mixed shear cracks, and multiple composite tensile-shear cracks that propagate along the bedding planes and penetrate the matrix. When $\theta = 90°$, the sample exhibits tensile and splitting cracks that propagate along its bedding planes, ultimately penetrating the entire sample.
(3) As the strain rate increases, the fracture modes of shale samples demonstrate notable differences. The fracture pattern of a 60° bedded specimen lacks consistency across various strain rates. Specifically, at a strain rate of $\dot{\varepsilon} = 20\ s^{-1}$, the fracture surface primarily extends along the bedding planes with shear cracks.

At $\dot{\varepsilon} = 50\ s^{-1}$, shear cracking prevails, accompanied by tensile cracks. As $\dot{\varepsilon}$ continues to increase, the fracture pattern becomes dominated by a combination of slip and tensile-shear mixed cracks.

(4) The loaded waveform has a substantial impact on the failure patterns of shale samples. As the peak stress intensity rises, the failure modes of the 30° and 60° bedded shale samples gradually shift from center-oriented failure to shear failure along the bedding planes. Meanwhile, the 90° bedded specimen initially exhibits a crack along the bedding planes. As the shock wave intensifies, secondary splitting cracks emerge within the matrix, indicating a fracture pattern that crosses the bedding planes.

References

Alhemdi, A., Gu, M.: Method to account for natural fracture induced elastic anisotropy in geomechanical characterization of shale gas reservoirs. J. Nat. Gas Sci. Eng. **101**, 104478 (2022)

Cai, X., Cheng, C., Zhou, Z., Konietzky, H., Song, Z., Wang, S.: Rock mass watering for rock-burst prevention: some thoughts on the mechanisms deduced from laboratory results. Bull. Eng. Geol. Env. **80**(11), 8725–8743 (2021)

Chao, Z., Shi, D., Zheng, J.: Experimental research on temperature-dependent dynamic interface interaction between marine coral sand and polymer layer. Ocean Eng. **297**, 117100 (2024)

Chen, B.P., Gong, B., Wang, S.Y., Tang, C.A.: Research on zonal disintegration characteristics and failure mechanisms of deep tunnel in jointed rock mass with strength reduction method. Mathematics **10**(6), 922 (2022)

Chen, T., Gong, B., Tang, C.: Origin and evolution of cracks in the glaze surface of a ceramic during the cooling process. Materials **16**, 5508 (2023)

Feng, X., Gong, B., Cheng, X., Chen, X., Xi, X., Wang, K. Investigation on dynamic anisotropy of bedded shale under SHPB impact compression. Eng. Fract. Mech. **320**, 111075 (2025)

Feng, X., Gong, B., Cheng, X., Zhang, H., Tang, C.: Anisotropy and microcrack-induced failure precursor of shales under dynamic splitting. Geomat. Nat. Haz. Risk **13**(1), 2864–2889 (2022)

Feng, X., Gong, B., Liang, Z., Wang, S., Tang, C., Li, H., Ma, T.: Study of the dynamic failure characteristics of anisotropic shales under impact Brazilian splitting. Rock Mech. Rock Eng. **57**, 2213–2230 (2024)

Gao, M., Gong, B., Liang, Z., Jia, S., Zou, J.: Investigation of the anisotropic mechanical response of layered shales. Energy Sci. Eng. **11**(12), 4737–4754 (2023)

Gong, B., Liang, Z.Z., Liu, X.X.: Nonlinear deformation and failure characteristics of horseshoe-shaped tunnel under varying principal stress direction. Arab. J. Geosci. **15**, 475 (2022)

Gong, B., Zhao, T., Thusyanthan, I., Tang, C.: Modelling rock fracturing by a novel implicit continuous to discontinuous method. Comput. Geotech. **166**, 106035 (2024)

Khmelinin, A.P., Konurin, A.I., Denisova, E.V.: Simulation of electromagnetic high-frequency wave propagation processes in multilayer geo-structures. Geohazard Mech. **1**(3), 203–207 (2023)

Kolawole, O., Oppong, F.: Assessment of inherent heterogeneity effect on continuous mechanical properties of shale via uniaxial compression and scratch test methods. Rock Mech. Bull. **2**(4), 100065 (2023)

Leroy, M.N.L., Joseph, N., Bienvenu, N.J.M.: Redistribution and magnitude of stresses around horse shoe and circular excavations opened in anisotropic rock. Int. J. Min. Sci. Technol. **25**(4), 615–621 (2015)

Li, Z., Li, L., Huang, B., Zhang, L., Li, M., Zuo, J., Li, A., Yu, Q.: Numerical investigation on the propagation behavior of hydraulic fractures in shale reservoir based on the DIP technique. J. Petrol. Sci. Eng. **154**, 302–314 (2017)

Li, D., Ma, J., Wan, Q., Zhu, Q., Han, Z.: Effect of thermal treatment on the fracture toughness and subcritical crack growth of granite in double-torsion test. Eng. Fract. Mech. **253**, 107903 (2021)

Liang, M., Wang, Z., Zhang, Y., Greenwell, C.H., Li, H., Yu, Y., Liu, S.: Experimental investigation on gas permeability in bedding shale with brittle and semi-brittle deformations under triaxial compression. J. Petrol. Sci. Eng. **196**, 108049 (2021)

Liu, H., Feng, X., Liu, L., Li, T., Tang, C.: Mechanical properties and failure characteristics of anisotropic shale with circular hole under combined dynamic and static loading. Int. J. Rock Mech. Min. Sci. **170**, 105524 (2023)

Liu, L., Zhang, Z., Wang, T., Zhi, S., Wang, J.: Evolution characteristics of fracture volume and acoustic emission entropy of monzogranite under cyclic loading. Geomech. Geophys. Geo-Energ. Geo-Resour. **10**, 16 (2024)

Luo, N., Fan, X., Cao, X., Zhai, C., Tao, H.T.: Dynamic mechanical properties and constitutive model of shale with different bedding under triaxial impact test. J. Petrol. Sci. Eng. **216**, 110758 (2022)

Luo, N., Zhang, H., Chai, Y., Li, P., Zhai, C., Zhou, J., Ma, T.: Research on damage failure mechanism and dynamic mechanical behavior of layered shale with different angles under confining pressure. Deep Undergr. Sci. Eng. **2**(4), 337–345 (2023a)

Luo, P., Li, D., Zhou, A., Ma, J., Zhu, Q., Jiang, J.: Influence of bedding plane on the tensile properties and crack propagation of soft and hard laminated rock-like under Brazilian test. Theoret. Appl. Fract. Mech. **127**, 104087 (2023b)

Ma, T., Wang, H., Liu, Y., Fu, C., Ranjith, P.G.: Experimental investigation on the anisotropy of mode-I fracture and tensile failure of layered shale. Eng. Fract. Mech. **290**, 109484 (2023)

Nguyen, T.S., Tanapalungkorn, W., Likitlersuang, S.: Probabilistic analysis of dual circular tunnels in rock masses considering rotated anisotropic random fields. Comput. Geotech. **170**, 106319 (2024)

Si, X., Luo, Y., Luo, S.: Influence of lithology and bedding orientation on failure behavior of "D" shaped tunnel. Theoret. Appl. Fract. Mech. **129**, 104219 (2024)

Tang, C.A., Webb, A.A.G., Moore, W.B., Wang, Y.Y., Ma, T.H., Chen, T.T.: Breaking earth's shell into a global plate network. Nat. Commun. **11**(1), 3621 (2020)

Tuzingila, R.M., Kong, L., Kasongo, R.K.: A review on experimental techniques and their applications in the effects of mineral content on geomechanical properties of reservoir shale rock. Rock Mech. Bull. **3**(2), 100110 (2024)

Wang, Y., Gong, B., Tang, C.: Numerical investigation on anisotropy and shape effect of mechanical properties of columnar jointed basalts containing transverse joints. Rock Mech. Rock Eng. **55**, 7191–7222 (2022a)

Wang, Y., Gong, B., Zhang, Y., Yang, X., Tang, C.: Progressive fracture behavior and acoustic emission release of CJBs affected by joint distance ratio. Mathematics. **10**, 4149 (2022b)

Wang, P., Ma, C., Liu, C., Liu, Q., Fu, Y., Cai, M.: Anisotropic behavior of excavated layered rock mass subjected to compression considering the joint roughness. Bull. Eng. Geol. Env. **82**(10), 373 (2023a)

Wang, P., Qi, Z., Ma, C., Cai, M.: Anisotropic behavior of the seepage-stress coupling mechanical model of coal pillars of underground reservoirs. Geomech. Geophys. Geo-Energ. Geo-Resour. **9**(1), 3 (2023b)

Wang, P., Ma, C., Zhang, B., Gou, Q., Tan, W., Cai, M.: Development of an improved three-dimensional rough discrete fracture network model: method and application. Int. J. Min. Sci. Technol. **33**(12), 1469–1485 (2023c)

Wang, Y., Chen, Q., Dai, B., Wang, D.: Guidance and review: Advancing mining technology for enhanced production and supply of strategic minerals in China. Green Smart Min. Eng. **1**(1), 2–11 (2024)

Wu, Y., Ma, D., Hu, X., Hao, Y., Liu, C., Zhou, H.: Numerical simulation on the mechanical and fracture behavior of bedding argillaceous sandstone containing two pre-existing flaws. Theoret. Appl. Fract. Mech. **127**, 104047 (2023)

Yu, Q., Liu, H., Yang, T., Liu, H.: 3D numerical study on fracture process of concrete with different ITZ properties using X-ray computerized tomography. Int. J. Solids Struct. **147**, 204–222 (2018)

Yu, C.Y., Gong, B., Wu, N., Xu, P.L., Bao, X.K.: Simulation of the fracturing process of inclusions embedded in rock matrix under compression. Appl. Sci. **12**(16), 8041 (2022)

Zhai, M., Xue, L., Bu, F., Yang, B., Huang, X., Liang, N., Ding, H.: Effects of bedding planes on progressive failure of shales under uniaxial compression: Insights from acoustic emission characteristics. Theoret. Appl. Fract. Mech. **119**, 103343 (2022)

Zhang, C., Zhu, Z., Wang, S., Ren, X., Shi, C.: Stress wave propagation and incompatible deformation mechanisms in rock discontinuity interfaces in deep-buried tunnels. Deep Undergr. Sci. Eng. **1**(1), 25–39 (2022)

Zhang, Y., Chen, S., Zhai, X., Khan, J., Zhang, Y.: Brazilian splitting experiment and finite element simulation analysis of the influence of bedding loading angle on shale fracture mode. Nat. Gas Ind. B **10**(6), 602–612 (2023)

Zhang, Y., Zhao, Y., Long, A., Wang, C., Bi, J.: Combined effects of bedding anisotropy and matrix heterogeneity on hydraulic fracturing of shales from Changning and Lushan, South China: an experimental investigation. J. Asian Earth Sci. **259**, 105908 (2024)

Zhao, C., Liu, J., Lyu, C., Liang, C., Xu, D., Ren, Y.: Mechanical responses and failure characteristics of longmaxi formation shale under real-time high temperature and high-stress coupling. Eng. Fail. Anal. **152**, 107490 (2023)

Zhou, X., Han, L., Bi, J., Shou, Y.: Experimental and numerical study on dynamic mechanical behaviors of shale under true triaxial compression at high strain rate. Int. J. Min. Sci. Technol. **34**(2), 149–165 (2024)

Zhu, Q., Li, D., Han, Z., Xiao, P., Li, B.: Failure characteristics of brittle rock containing two rectangular holes under uniaxial compression and coupled static-dynamic loads. Acta Geotech. **17**, 131–152 (2022)

Chapter 4
Numerical Simulation on Dynamic Splitting of Brazilian Shale Disc

Abstract The influence of bedding planes on the dynamic anisotropy and failure mechanisms of shales was investigated through a series of Brazilian disc tests using the split Hopkinson pressure bar system. These tests were conducted to understand how beddings affect the anisotropic strength and deformation characteristics of shales under dynamic impact. The fracturing process and failure patterns were observed using a high-speed camera. Additionally, to comprehend the fracture initiation and propagation mechanisms, stress buildup, stress shadow, and stress transfer were modeled using digital image processing and rock failure process analysis. The experimental and numerical results were consistent, allowing for an analysis of the effects of dip angle and spacing interval of beddings on crack initiation, propagation, and coalescence. Furthermore, numerical simulations captured the spatial location and energy magnitude of crack-induced acoustic emissions, providing insights into the energy release patterns during dynamic Brazilian splitting. The findings reveal significant differences in failure characteristics and mechanical mechanisms along different load directions. As the dip angle of the beddings increases, the area percentage of cracked beddings decreases, while the area percentage of cracked rock matrix increases. For a constant dip angle, the area percentages of cracked beddings and rock matrix fluctuate irregularly with increasing bedding spacing. These findings significantly enhance our understanding of the dynamic anisotropy and failure mechanisms of bedded shales, contributing to improved dynamic excavation and support design for layered rock masses.

Keywords Dynamic anisotropy · Split Hopkinson pressure bar · Brazilian splitting · Bedded shale · Acoustic emission

4.1 Introduction

The stability and safety of rock masses are significantly impacted by naturally occurring discontinuities. Shales, which are abundant on Earth and often encountered in slope excavation, tunneling, mining, and other engineering activities (Li et al. 2020;

© The Author(s), under exclusive license to Springer Nature Singapore Pte Ltd. 2025
X. Feng and B. Gong, *Mechanical Properties of Layered Rock Mass*,
https://doi.org/10.1007/978-981-96-5142-9_4

Zhang et al. 2019; Yu et al. 2019; Xu et al. 2019), frequently exhibit a distinctive bedding structure due to geological processes such as weathering, transportation, and sedimentation (Zhang et al. 2018). The mechanical properties of bedding planes differ significantly from those of the rock matrix, leading to strong anisotropy in shales under dynamic impact. This anisotropy can result in uneven deformation and unexpected failures in engineered rock masses, posing risks to life and property (Wang et al. 2020a; Zhang et al. 2017; Ma et al. 2017).

The dynamic anisotropy of shales has a profound impact on various strength and deformation properties, depending on their structural characteristics and load/unloading directions (Suo et al. 2020; Wang et al. 2016; Khanlari et al. 2014; Eliyahu et al. 2015). Under dynamic loads, micro-cracks may form at pre-existing defects in the shale matrix, such as voids, grain boundaries, and inclusions (Harris et al. 2011). Moreover, the bedding structure, a key characteristic of shales, plays a crucial role in governing their nonlinear mechanical response under dynamic disturbance and is essential in shale-related projects (Wang et al. 2020c).

To investigate dynamic failure mechanisms, many researchers have utilized the split Hopkinson pressure bar (SHPB) system, which is closely related to stress wave propagation in heterogeneous rock (Shang et al. 2016; Dai et al. 2016). Wang et al. (2020b) employed notched semi-circular bend (NSCB) specimens to study the influence of bedding surfaces on crack propagation characteristics. Vervoort et al. (2014) proposed a function linking the inclination angle of weak planes to strength variations and final fracture patterns based on experimental data. Guha Roy and Singh (2016) found that both heat treatment and layer orientation strongly control tensile fracture geometry. Huang et al. (2020a, b) indicated that continental shales exhibit strong anisotropy in fracture evolution and failure mode, with the uniaxial compressive strength and degree of fracture exploiting beddings showing an inverse trend as the bedding angle increases from 0° to 90°. Ma et al. (2018) classified failure modes of anisotropic rocks into five categories, including tensile and shear failures across and along weakness planes, and mixed failures. Fan et al. (2021) observed post-peak strain softening in shale samples. Sun et al. (2019) suggested that splitting tensile failure is more likely at low bedding inclinations, while shear or mixed failure is more prone at medium inclinations. These studies have significantly contributed to our understanding of the dynamic behavior of bedded shales and their failure mechanisms, providing valuable insights for engineering practices involving layered rock masses.

In addition to focusing on anisotropic mechanical strength and eventual failure morphology, researchers have also directed significant attention to the fracturing process and failure mode (Huang et al. 2020a, b). Tien et al. (2006) utilized a rotary scanner to capture "unrolled" images of rock specimens during uniaxial compressive tests at various stress levels, aiding in the examination of failure processes and modes. Ma and Chen (2014) analyzed the relationship between shale hydration meso-damage and soaking time by segmenting CT images. Wu et al. (2016) emphasized the crucial role of meso-scale structure in influencing the failure process of samples under different loading conditions. However, due to monitoring limitations, the mechanical mechanisms underlying stress buildup, stress shadow, and stress transfer (SSS) processes during the creation and propagation of fractures in bedded shales under dynamic impact remain elusive. Furthermore, the intricate dynamic failure behaviors of heterogeneous rock masses containing mesoscopic pre-existing defects and macroscopic structural planes necessitate further investigation.

Moreover, the failure of transversely isotropic rock often exhibits greater complexity than presumed (Duan and Kwok 2015; Dan et al. 2013). Numerical technology has emerged as a potent tool to deepen our understanding of failure mechanisms in anisotropic and inhomogeneous materials, complementing laboratory tests (Bahaaddini et al. 2013). Duan et al. (2016) and Liu et al. (2019) modeled the mechanical behaviors of inherently anisotropic rocks within a specific angle range using smooth-joint contacts based on the discrete element method (DEM). Wang et al. (2018) employed discrete fracture network (DFN) models with diverse joint distributions to study the directionality of jointed rock mass in alignment with laboratory test configurations. However, DEM faces challenges in estimating contact displacements between blocks/particles, ensuring algorithm convergence, and obtaining accurate deformation fields (Brara et al. 2001; Tang et al. 2000). Additionally, it struggles to efficiently reproduce detailed fracturing processes. Zhu and Tang (2006) introduced an elastic damage-based law that considers strain-rate dependency to describe the constitutive law using the rock failure process analysis (RFPA) method (Liang et al. 2012; Qian et al. 2020). Meng et al. (2021) established a cohesive zone model to investigate crack distributions influenced by the number and strength of bedding planes. Wu et al. (2020) observed that shale fractal characteristics become more complex as osmotic pressure increases, using a seepage-stress coupling model. Huang et al. (2020a, b) analyzed the impact of bedding parameters on fracture length percentage along the bedding plane and the proportion of tensile and shear cracks. Shang et al. (2018) discussed the effects of bedding lanes with varying relative positions and spacings. Xu et al. (2018) noted that fracture surfaces tend to form roughly parallel to the load direction at the center of rock disks with loose structures or strong foliation. However, conventional numerical methods encounter difficulties in addressing the natural heterogeneity and pre-existing discontinuities of jointed rock masses and modeling the intricate dynamic failure process involving crack creation, growth, and coalescence under dynamic load conditions.

In this paper, a split Hopkinson pressure bar system was employed to conduct a series of dynamic Brazilian splitting tests on shale samples. Subsequently, the study examined the variations in tensile strengths and failure patterns of these samples with respect to different bedding dip angles. Utilizing digital image processing (DIP) technology and the RFPA method, corresponding numerical models were developed. The accuracy of these numerical models was confirmed by the strong correlation between experimental and simulated results. Further investigation delved into the influence of dip angle and bedding spacing interval on crack initiation, propagation, and coalescence within Brazilian disks subjected to dynamic loading. Additionally, the spatial distribution and energy magnitude of crack-induced acoustic emissions (AEs) were discussed, providing insights into the nonlinear deformation and failure characteristics of bedded shales. These findings are expected to contribute significantly to stability analysis, support design, and disaster prevention and mitigation strategies in rock engineering projects involving bedded shales.

4.2 Material and Method

4.2.1 Preparation of Shale Samples

To explore the anisotropic characteristics, shale samples were obtained from outcrops in Jiangsu Province, China, for this study. Cylindrical specimens, each with a diameter of 75 mm and a thickness of 30 mm, were drilled from the shale formations and prepared for testing in accordance with the guidelines provided by the International Society for Rock Mechanics (ISRM) (Aminzadeh et al. 2019). These shale samples are characterized by alternating thin, weak beddings and thick, strong layers, exhibiting a transversely isotropic nature. Notably, the beddings/layers are nearly parallel to each other.

4.2.2 Test Procedure

The SHPB system, a well-established method for assessing the dynamic response of rocks, was utilized in this study. The impact system comprised a gas gun, a striker bar, an incident bar, a transmission bar, and an absorption bar. Specifically, the striker, incident, and transmitted bars measured 200 mm, 3000 mm, and 1800 mm in length, respectively. These cylindrical bars, along with the striker, were constructed from hardened 60Si2Mn material, possessing a density of 7850 kg/m^3. The bars featured a diameter of 38.1 mm, a Young's modulus of 211 GPa, and a P-wave velocity of 5102 m/s. To ensure a gentle loading rise and achieve dynamic equilibrium in the brittle shale samples prior to failure, a pulse shaping technique was employed. This involved the use of a copper disc with a diameter of 3 mm and a thickness of 0.6 mm as the pulse shaper. Strain gauges and a dynamic strain amplifier were utilized to measure the strain in the incident and transmitted bars. Data analysis and storage were facilitated by an oscilloscope with a sampling rate of 10 MHz, interfaced with a computer. Furthermore, a high-speed camera was employed to capture the progressive failure process during impact loading. Images were captured at an interval of 10 μs, providing detailed insights into the failure mechanisms of the shale samples.

When an impact wave travels through the bars, they undergo strain. The histories of the incident strain (ε_i), reflected strain (ε_r), and transmitted strain (ε_t) were captured by strain gauges installed on the bars. To determine the displacement and load in the specimen, the three-wave analysis method was employed, as outlined by Lifshitz and Leber (1994) and Sharafisafa et al. (2020):

$$P_1(t) = P_2(t) \tag{4.1}$$

where P_1 is the force on the incident end of the sample, and P_2 on the transmitted end of the sample. They can be can be calculated as follows:

$$P_1(t) = A_0 E[\varepsilon_I(t) + \varepsilon_T(t)] \tag{4.2}$$

$$P_2(t) = A_0 E \varepsilon_R(t) \tag{4.3}$$

where ε_I, ε_R and ε_T represent the dynamic strain of the incident, reflected and transmitted waves, respectively; E is the Young's modulus of the bar materia; A_0 is the cross-sectional area of the bar.

4.3 Experimental Results and Analysis

Figure 4.1 illustrates the sequential impact failure process of specimens with varying bedding dip angles under a low strain rate of approximately 40 s^{-1}. Distinct characteristics in crack initiation, propagation, main crack formation, and macroscopic fracture patterns are observed among shale discs with different bedding dip angles. Initially, cracks emerge primarily within the first 90 μs. These initial cracks are barely visible due to their minute crack openings. Sustained loading leads to the formation of microcracks at the weakest bedding planes. Concurrently, tensile strain localization occurs at the specimen's end adjacent to the incident bar, extending along the wave propagation direction. As the loading time increases between 90 and 190 μs, the main cracks propagate along distinct paths influenced by the bedding surfaces. Specifically: For a bedding angle θ of 0°, the primary vertical crack traverses the specimen's center and generally aligns with the bedding surfaces. Additionally, localized vertical cracks near the loading jaw contribute to the formation of a compressed zone. When the bedding angle θ is 30°, the main crack adopts a more intricate circular arc shape, reflecting the interplay between vertical loading and inclined bedding planes.

As the bedding angle θ rises to 60°, a primary crack aligns with the loading direction, accompanied by a secondary, shorter crack parallel to it. When θ reaches 90°, a vertical main crack traverses the specimen's centerline, perpendicular to the bedding plane. Adjacent to this main crack, a secondary tensile crack arises due to tensile stress concentration at the relatively weaker bedding plane. As the dynamic impact progresses to the final stage of macroscopic failure, the crack openings widen, revealing diverse fracturing patterns: along the bedding surface, at the disc's center, and along curved paths. Furthermore, the crack initiation mechanism is evident at points of weakness. The disc continues to produce parallel and curved cracks due to high stresses at the tips of initial cracks. Ultimately, a macro-intersect crack forms through the gradual coalescence of pre-existing cracks.

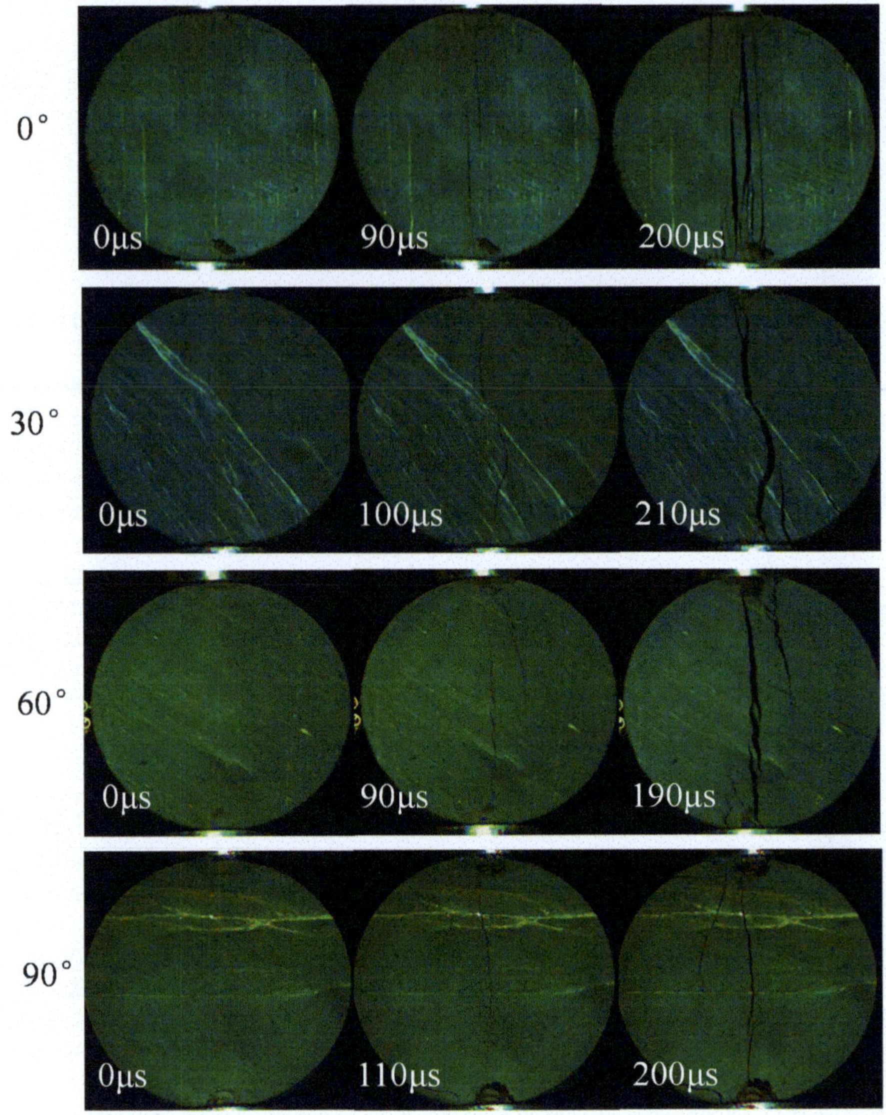

Fig. 4.1 Failure features of the shale discs shot. Modified after Feng et al. (2022)

The generation and propagation of microcracks stem from both the damage to the bedding planes and tensile failure originating at the disc's center. The bedding structure's influence can alter the cracking direction, potentially resulting in the formation of arc-shaped cracks. Moreover, these cracks can coalesce and interconnect across various bedding layers, ultimately culminating in macroscopic fracture. Consequently, this failure process is notably influenced by the anisotropy and heterogeneity of shale rocks.

Figure 4.2 illustrates the trend of nominal tensile strengths of shale discs during dynamic Brazilian splitting in polar coordinates. The strength varies in a U-shaped manner as the bedding angle increases. Initially, the tensile strength diminishes, followed by an increase as the bedding angle grows. Comparing the dynamic strength of shales with the quasi-static strength reported by Khanlari et al. (2015), we observe a striking similarity in their trend as the bedding angle rises, both exhibiting a U-shaped pattern. When the bedding angle θ approximates 30°, the tensile strength reaches its minimum. Subsequently, as θ increases from 30° to 90°, the tensile strength augments. The maximum tensile strength of shales is attained when the bedding angle θ equals 90°. Hence, the strength values are intricately linked to the internal structural surfaces of shales.

To quantify the impact of bedding planes, the crack distributions within both the bedding planes and the rock matrix were further categorized. Figure 4.3a exhibits the length variation of two types of fractures—those occurring within the bedding planes and those within the rock matrix—as a function of the shale's bedding angle θ. It reveals that as the bedding angle θ increases, the fracture length within the rock matrix also increases. Meanwhile, Fig. 4.3b highlights the significant influence of the bedding angle θ on the cracking percentage, particularly when θ transitions from 0° to 30°. Beyond $\theta = 30°$, the cracking percentages in both the bedding planes and

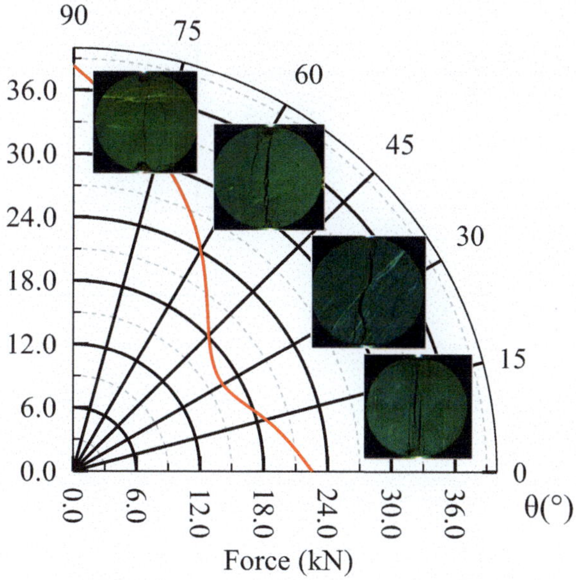

Fig. 4.2 Nominal tensile strength of Brazilian discs. Modified after Feng et al. (2022)

the rock matrix fluctuate. Notably, the fracture length and cracking percentage within the bedding planes exhibit a similar trend as the bedding angle θ increases.

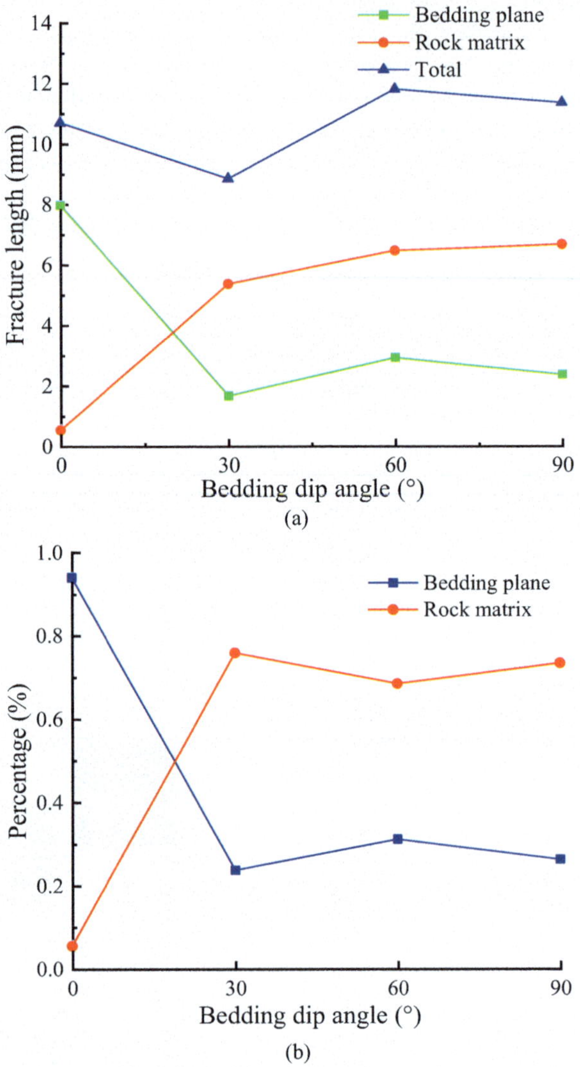

Fig. 4.3 **a** Average fracture length changing with the bedding angle θ of shale and **b** cracking percentage of rock matrix and bedding planes changing with the bedding angle θ of shale (Feng et al. 2022)

4.4 Numerical Simulation

4.4.1 *Principle of Numerical Procedure*

To gain deeper insights into the mechanical mechanisms driving the failure characteristics of shales, we integrated the RFPA (Realistic Failure Process Analysis) method with digital image processing (DIP) to capture stress/strain variations and reproduce the gradual failure process of shale discs refer to Sect. 3.2.

4.4.2 *Model Establishment*

In this study, the numerical discs had a diameter of 75 mm. The image resolution was set to 470 × 660 pixels, and the model comprised a total of 310,200 elements. The physical and mechanical parameters utilized in the simulations are outlined in Table 4.1. The bedding dip angles considered were $\theta = 0°$, 30°, 60°, and 90°, consistent with previous tests. Additionally, to investigate the impact of bedding spacing on the mechanical response of shales, three spacing distances were analyzed: $S_{bp} = 1.5$ mm, 2.75 mm, and 4 mm, respectively, with all other parameters held constant. The incident compressive stress wave, simplified from the aforementioned physical experiment, was applied to the top loading plate, as illustrated in Fig. 4.4. Based on the plane strain hypothesis, the gradual splitting process of the shale discs was simulated.

4.4.3 *Numerical Results and Analysis*

4.4.3.1 Progressive Brazilian Splitting

Figure 4.5 depicts the continuous failure process of the sample under dynamic impact. It is evident that the stress wave propagation within the sample evolves concurrently with the damage to the elements. Prior to the stress wave reaching the bottom of

Table 4.1 Physical and mechanical parameters used in simulation (Feng et al. 2022)

Parameter	Bedding plane	Matrix	Steel plate
Density (kg/m^3)	2500	2500	7800
Young's modulus (GPa)	5	50	210
Poisson's ratio	0.2	0.2	0.2
Homogeneity index *m*	3	8	100
Compressive strength (MPa)	50	100	500

Fig. 4.4 Incident wave used in the numerical simulation (Feng et al. 2022)

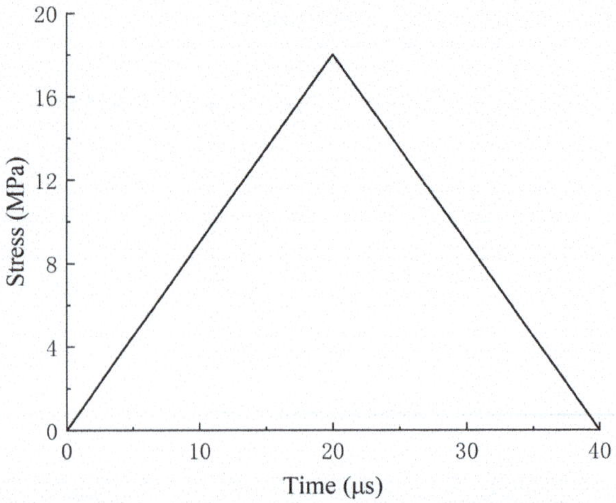

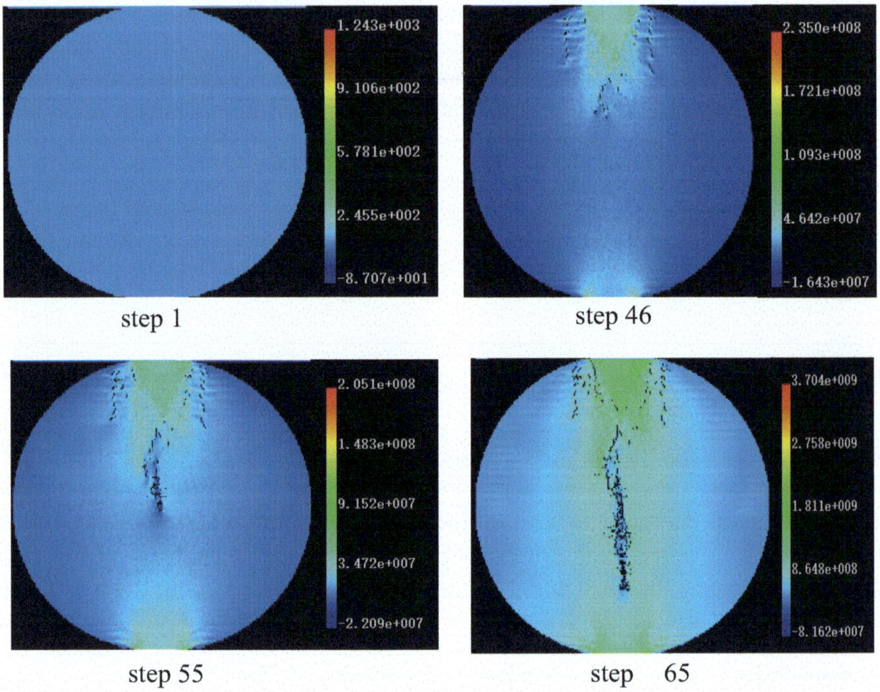

Fig. 4.5 The progressive splitting of the disc by the maximum principal stress contours. Modified after Feng et al. (2022)

the sample, only a limited number of elements at the top region experience damage. As the loading continues, cracks initiate near the upper loading plate due to the high compressive stress. Subsequently, these cracks propagate downward towards the lower part of the sample, resulting in multiple visible cracks. Under further dynamic loading, an increasing number of elements in the central part of the sample undergo damage, forming a tensile fracture zone. Ultimately, a macroscopic fractured zone comprising multiple cracks is established along the vertical diameter of the disc. The simulated fracture characteristics and failure mode align well with experimental observations, demonstrating the effective modeling of the progressive Brazilian splitting process.

4.4.3.2 Crack Propagation of Brazilian Discs with Different Bedding Spacings

In this section, we delve into the impact of bedding dip angle and bedding spacing on the nonlinear deformation and failure behaviors of Brazilian discs. Figure 4.6 presents the modeled fracture characteristics and failure patterns, illustrated through the maximum principal stress contours of the specimens.

When the bedding dip angle θ is 0°, cracks emerge along the vertical bedding surfaces in the middle of the sample, aligning with experimental results as shown in Fig. 4.4. The sample splits into two large pieces along a direction parallel to the loading axis. However, the effect of bedding spacing on the failure mode is relatively minor at $\theta = 0°$. As the bedding dip angle θ increases to 30° and the bedding spacing S_{bp} is 1.5 mm, a main crack forms at the center of the sample, accompanied by a bending inflection point at the middle right part due to the inclined bedding surfaces. This mixed-mode fracture phenomenon also aligns with experimental observations, validating the reliability and effectiveness of the simulation approach. Furthermore, as the bedding spacing increases to 2.75 mm and 4 mm at $\theta = 0°$, single layer activation and central fracture occur respectively. When $\theta = 60°$, the main fracture occurs at the center of the samples with $S_{bp} = 1.5$ mm, 2.75 mm, and 4 mm. However, numerous secondary cracks appear in the central fractured zone due to high stress concentrations at the inclined bedding surfaces, consistent with experimental results. At $\theta = 90°$, a prominent central splitting forms regardless of the bedding spacing distance.

Figure 4.7a–c illustrate the variation in cracking percentage of both the bedding planes and the rock matrix across three different bedding spacings. Generally, as the bedding dip angle increases, the cracking percentage of the bedding surfaces decreases while the cracking percentage of the rock matrix increases. However, the specific patterns of these changes are influenced by the bedding spacing. From Fig. 4.7d, it can be observed that for a given bedding dip angle, the changes in cracking percentage of both the bedding surfaces and the rock matrix are quite intricate. When the bedding spacing S_{bp} is 1.5 mm, the cracking percentage of the rock matrix significantly rises as θ increases from 0° to 30°, and then remains relatively stable as θ changes from 30° to 90°. Furthermore, when the bedding spacing S_{bp} is 4 mm,

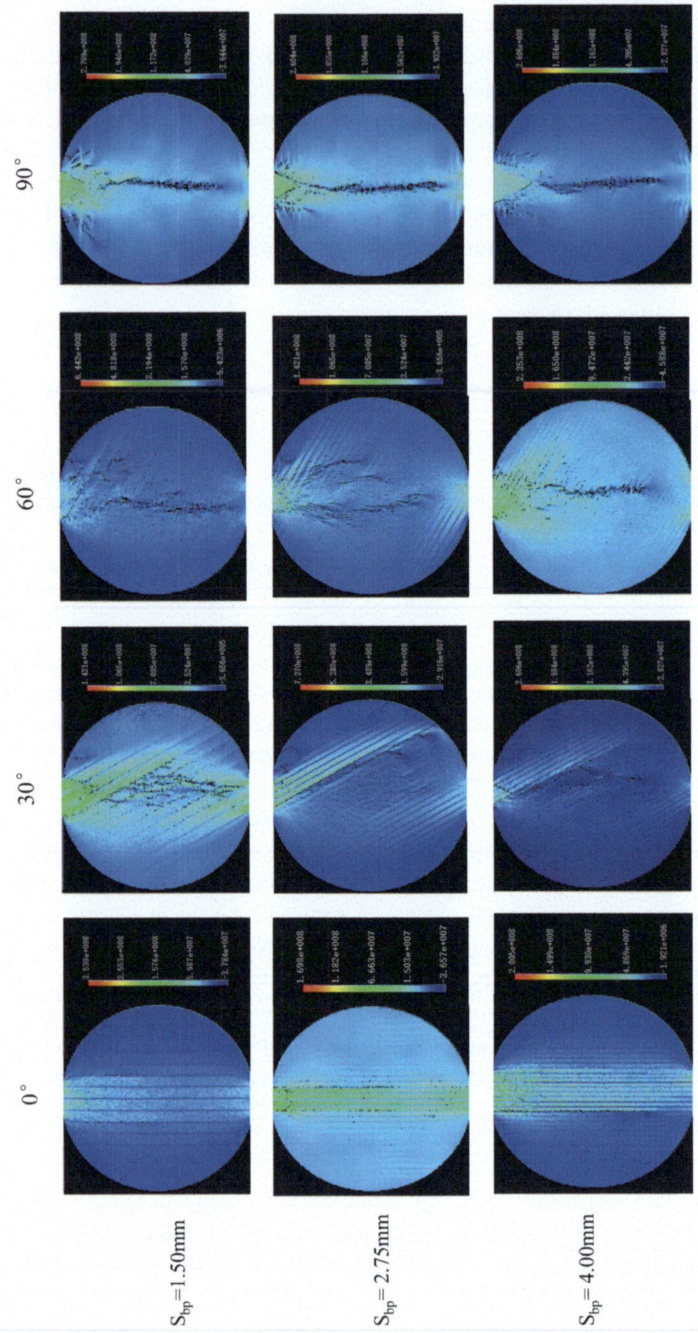

Fig. 4.6 Modeled fracture characteristics and failure patterns. Modified after Feng et al. (2022)

the cracking percentage of the rock matrix increases steeply as θ increases from 0° to 60°, and then plateaus as θ changes from 60° to 90°. However, when the bedding spacing S_{bp} is 2.75 mm, the cracking percentage of the rock matrix continues to rise as θ increases from 0° to 90°. Thus, the bedding spacing exerts a significant and complex influence on crack distribution.

The fracture force curves of Brazilian discs with varying bedding spacings all exhibit a U-shaped pattern as the bedding angle increases, as depicted in Fig. 4.8.

Fig. 4.7 Cracking percentage of rock matrix and bedding planes. Modified after Feng et al. (2022)

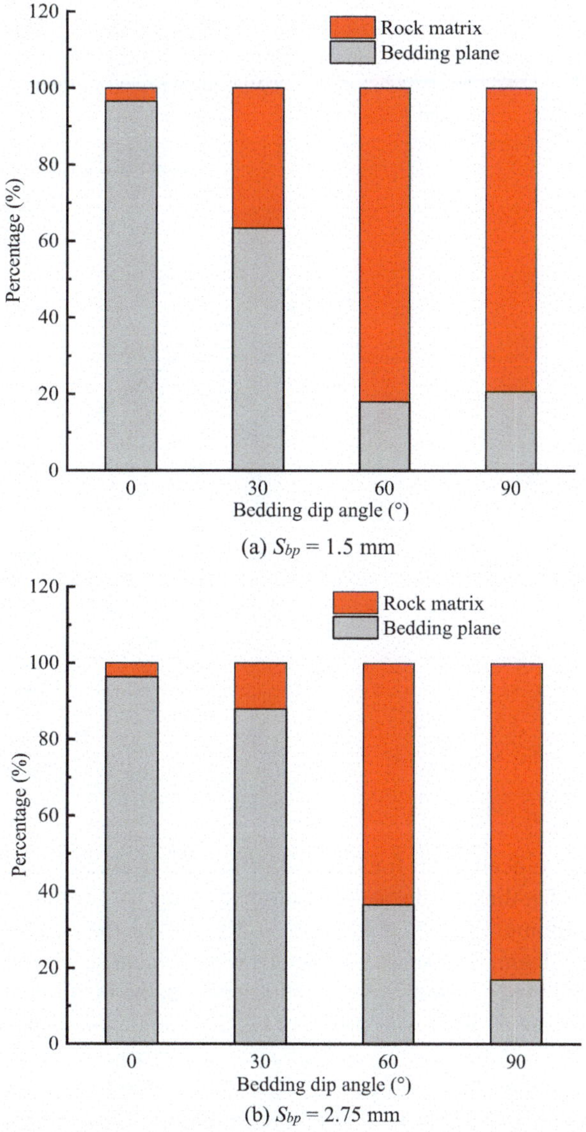

Fig. 4.7 (continued)

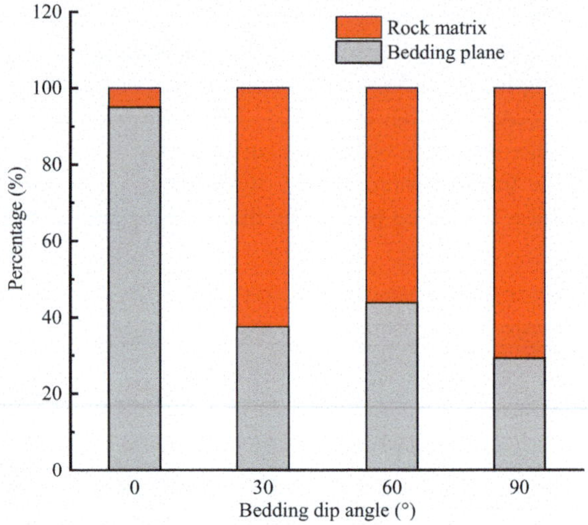

(c) $S_{bp} = 4$ mm

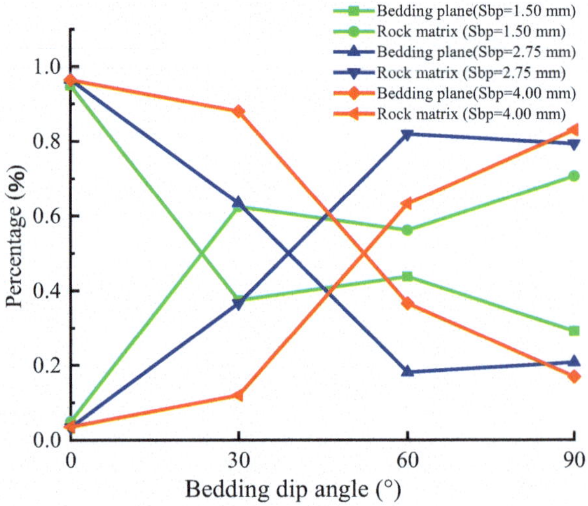

(d) Effect of bedding planes

Specifically, the fracture force initially decreases and then increases with the rise in bedding angle. At a bedding spacing of 1.5 mm, the fracture force reaches its minimum of 45.08 kN at $\theta = 30°$. When the bedding spacing is either 2.75 mm or 4 mm, the smallest fracture forces occur at $\theta = 60°$, measuring 51.02 kN and 55.44 kN, respectively. Additionally, as the bedding spacing increases, the fracture forces of discs with different bedding angles increase in diverse manners. Notably, at a bedding spacing of 1.25 mm, the fracture force increases nearly linearly at $\theta = 0°$; as the bedding spacing rises from 1.25 mm to 4 mm, the fracture force increases from 45.02 kN to 57.67 kN, representing an approximate 28% increase at $\theta = 30°$.

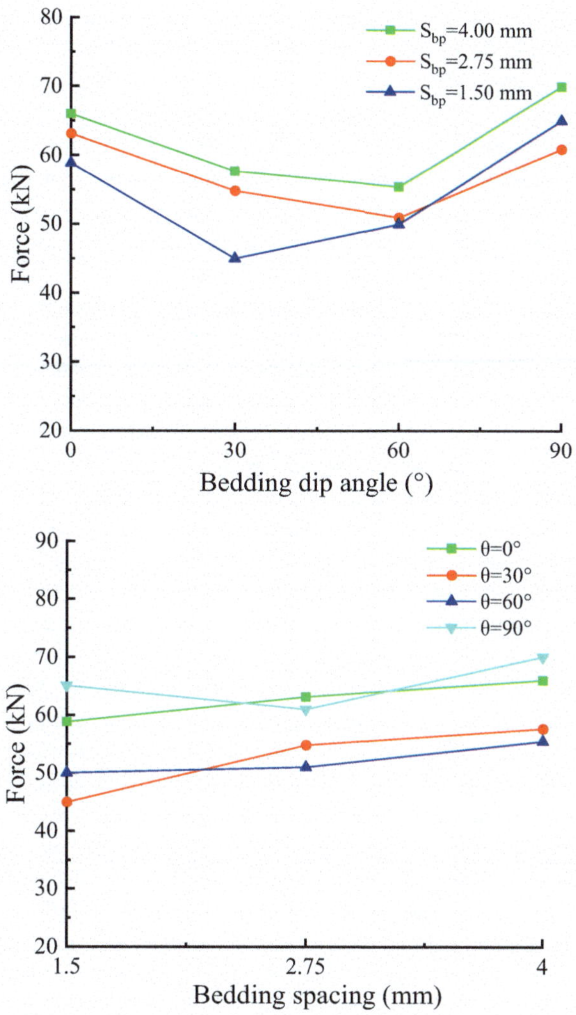

Fig. 4.8 Effect of bedding dip angle and bedding spacing on fracture force. Modified after Feng et al. (2022)

At $\theta = 60^\circ$ and 90°, this increment in fracture force is about 5.44 kN and 4.94 kN, respectively. Therefore, it is evident that the fracture force is significantly influenced by the interplay between bedding spacing and bedding angle.

4.4.3.3 Acoustic Emission Analysis

When a rock-like quasi-brittle material undergoes damage, acoustic emissions (AEs) are released in the form of elastic waves. The generation of these transient elastic waves signifies the rapid release of energy within a localized area. To explore the spatiotemporal evolution of elastic energy in shales resulting from crack initiation and growth under dynamic impact, this section discusses the spatial distribution and magnitude of AE events. Figures 4.9, 4.10, 4.11, 4.12, 4.13 and 4.14 depict the simulated AE performance during the failure process of shale discs. These figures illustrate that each AE event corresponds to the failure of an element within the shale model, and the number of AE events is directly proportional to the number of damaged elements. It is important to note that red circles represent AE events resulting from tensile failure of an element, while white circles represent AE events stemming from compression-shear failure. The radius of each circle is proportional to the magnitude of the released energy.

The first row of Fig. 4.9 illustrates that a few elements are initially damaged due to the compression-shear strength criterion being met, with several AE events releasing relatively large amounts of energy near the incident plate during the initial stage of dynamic loading. As the load stress increases, micro-cracks form at the weakest parts of the shale disc. Notably, when the bedding angle $\theta = 90°$, compression failures occur prominently along the bedding planes at the top area of the sample, forming a series of parallel rupture bands. As the dynamic stress wave continues to propagate downwards, more and more meso-elements reach their strength thresholds, and the pre-existing cracks propagate and evolve primarily along the vertical or inclined bedding planes when the bedding angle θ is 0°, 30°, or 90°. However, for $\theta = 90°$, a vertical broken zone spanning the bedding surfaces is induced by the high tensile stress beneath the previously formed horizontal compression bands. The fourth row of Fig. 4.9 displays the spatial positions of all AE events when the final failure modes develop, revealing that tensile damage plays a crucial role in triggering the macroscopic failure of bedded shales under splitting conditions.

Furthermore, Fig. 4.10 presents the corresponding AE counts and accumulated AE energy during the Brazilian splitting test. At the initial stage of dynamic loading, the number of AE events increases slightly, and the cumulative energy curves rise gradually, exhibiting an overall upward trend. As cracks evolve and develop, the AE counts begin to escalate. At the onset of macroscopic failure, the AE count curves steepen significantly, indicating that the samples begin to generate a substantial number of AE signals. Simultaneously, the cumulative energy reaches its maximum value very rapidly. Therefore, the acceleration in crack growth and the rapid increase in AE energy can be considered reliable precursors of shale failure. With the coalescence and interconnection of pre-existing cracks, the primary macroscopic fracture gradually forms, and the cumulative AE energy attains its peak value. Subsequently, although AE activities diminish, AE events continue to occur, suggesting that the specimens retain residual strength and that some elements are still undergoing damage.

Figure 4.11 displays the AE spatial evolution characteristics of shale discs with varying bedding dip angles when the bedding spacing is densified to 2.75 mm. The first row demonstrates that, at the inception of dynamic loading, several AE events arise due to damage near the incident plate. Notably, when the bedding angle $\theta = 0°$, compression-shear damage dominates. As the load stress intensifies, cracks propagate downwards for $\theta = 0°$, 30°, and 60°, but their pathways are significantly influenced by the presence of bedding surfaces. Conversely, for $\theta = 90°$, two small

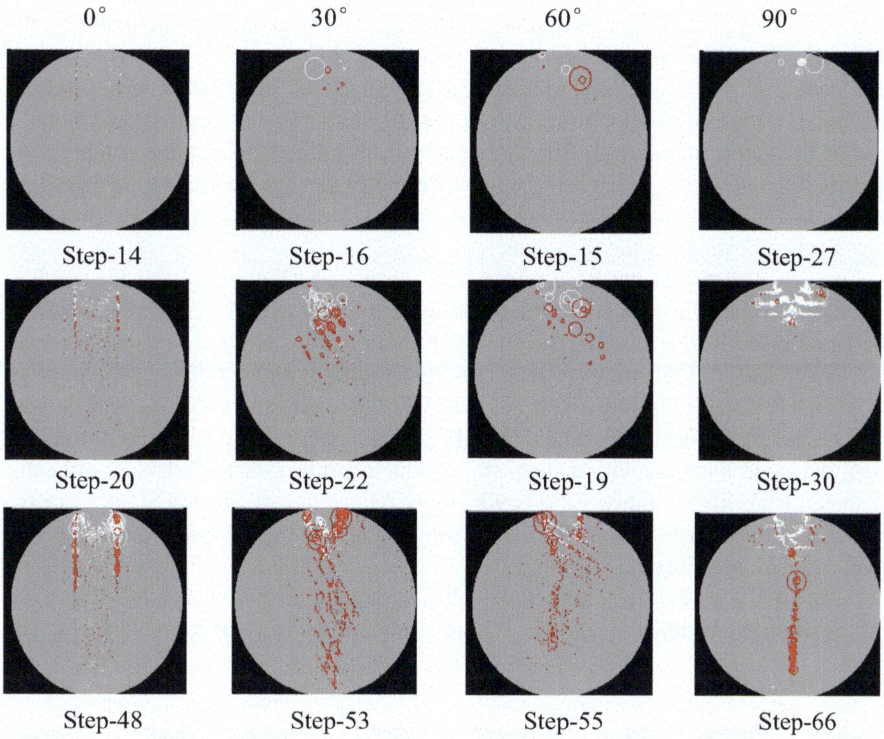

Fig. 4.9 AE distribution when $S_{bp} = 4$ mm. Modified after Feng et al. (2022)

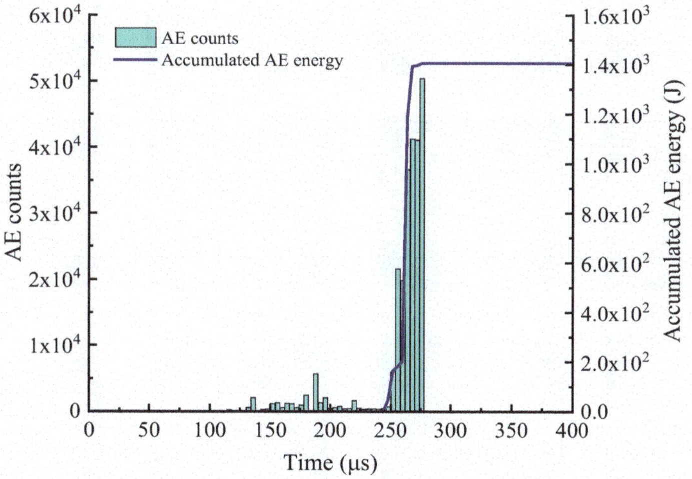

Fig. 4.10 AE events and energy of 0° sample when $S_{bp} = 4$ mm. Modified after Feng et al. (2022)

compression rupture zones form beneath the ends of the upper loading plate. As the dynamic stress wave progresses, cracks propagate along the bedding surfaces for $\theta = 0°$ and 30°. However, as θ increases to 60°, the rock located between adjacent beddings at the middle-upper section of the disc experiences more severe damage due to the steep inclination. Simultaneously, a prominent triangular compression rupture zone emerges beneath the upper loading plate. The fourth row of Fig. 4.11 illustrates that elements are primarily damaged in tensile mode due to high tensile stress concentrations at the disc center.

Figure 4.12 displays the AE counts and accumulated AE energy when the bedding spacing (S_{bp}) is 2.75 mm. It is evident from the figure that at the commencement of the Brazilian splitting test, AE events occur sporadically, and there is no significant energy release through these events. As cracks initiate and propagate, the number of AE events gradually increases, but the cumulative AE energy remains relatively low. As macro-failure approaches, the AE counts and cumulative AE energy curves exhibit a rapid increase with a steep slope, which can be considered as an indicator of shale instability. Notably, for $\theta = 90°$, more AE events are generated after macro-failure, but the energy released is small, suggesting that the sample retains a certain level of residual strength.

Figure 4.13 depicts the AE spatial evolution of samples with varying bedding dip angles when the bedding spacing is 1.5 mm, indicating denser bedding surfaces. The

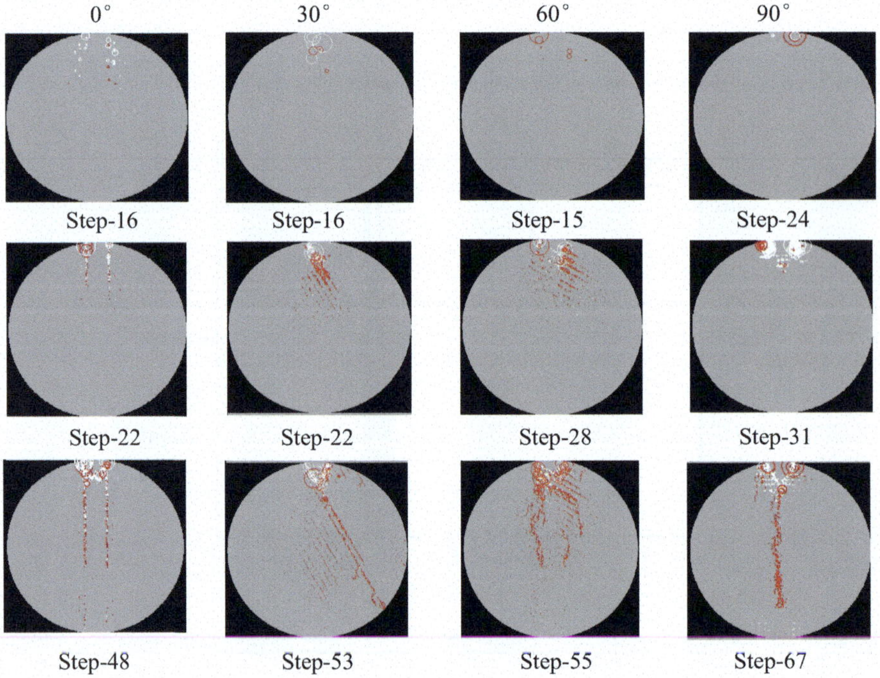

Fig. 4.11 AE distribution of shale discs when $S_{bp} = 2.75$ mm. Modified after Feng et al. (2022)

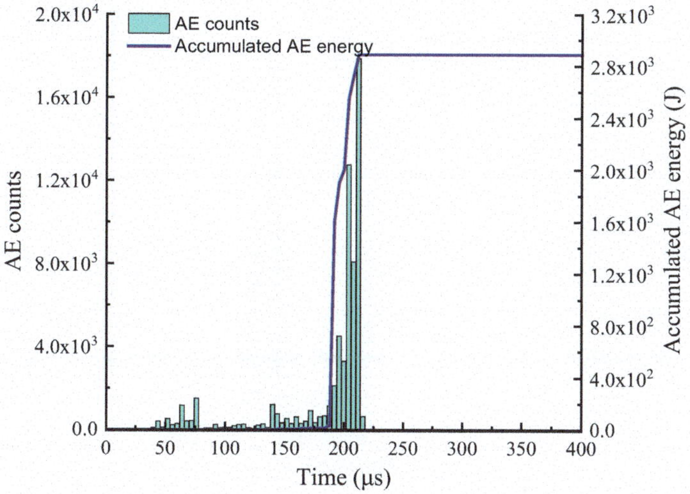

Fig. 4.12 AE events and energy of 0° sample when $S_{bp} = 2.75$ mm. Modified after Feng et al. (2022)

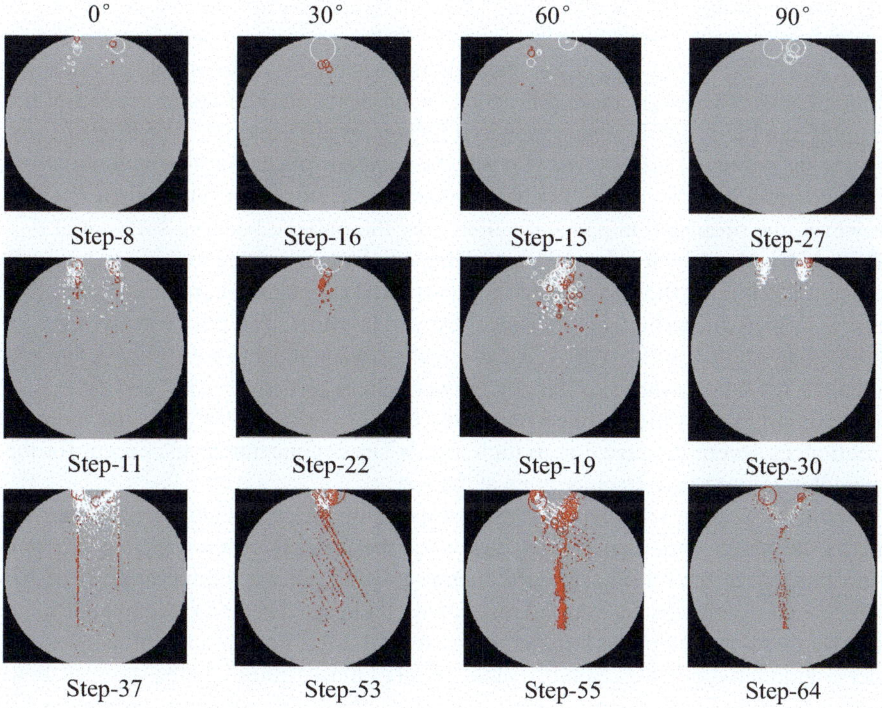

Fig. 4.13 AE distribution of shale discs when $S_{bp} = 1.5$ mm. Modified after Feng et al. (2022)

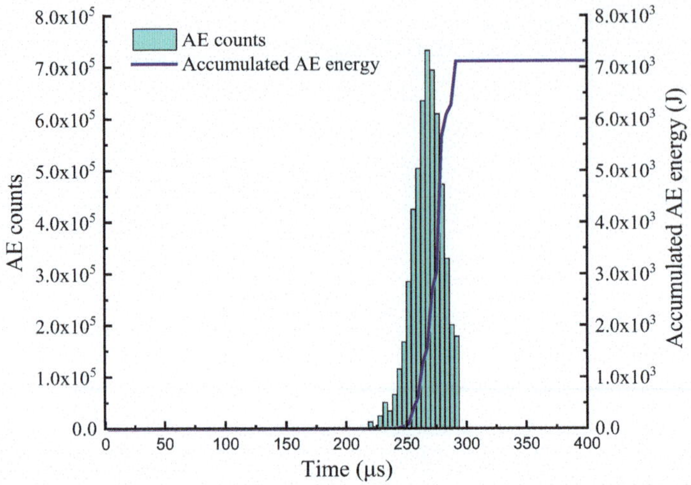

Fig. 4.14 AE events and energy of 0° sample when $S_{bp} = 1.5$ mm. Modified after Feng et al. (2022)

first row illustrates that, at the initial stage of dynamic impact, several AE events concentrate near the incident plate, similar to the previous scenarios with $S_{bp} =$ 4 mm and 2.75 mm. Compression-shear damage is predominant in these events. As the load stress increases, more AE events occur at the middle-upper section of the samples, but their distribution is scattered for $\theta = 0°$, 30°, and 60°. For $\theta = 90°$, two significant damaged areas form beneath the incident plate. As the dynamic stress wave continues to propagate, cracks propagate straight downwards when $\theta = 0°$. However, the propagation paths are altered by the oblique bedding surfaces at $\theta =$ 30° and 60°. Notably, for $\theta = 60°$, a bending crack emerges due to damage between adjacent beddings at the disc's center. Concurrently, a triangular compression fracture zone is formed due to high compressive stress in this area. The third row of Fig. 4.13 shows that when $\theta = 0°$, two vertical tensile cracks appear, with many elements damaged in compression-shear mode between them. When $\theta = 30°$, tensile cracks primarily align with the inclined beddings. For $\theta = 60°$ and 90°, the disc's center experiences severe damage due to high tensile stress concentrations, except for the triangular compressive fracture zone at the top.

Furthermore, Fig. 4.14 displays the corresponding AE counts and accumulated AE energy during the Brazilian splitting test when the bedding spacing (S_{bp}) is 1.5 mm. It reveals that at the initial stage of dynamic loading, only a limited number of AE events occur, with very low cumulative released energy. However, the overall trend is upward. As more cracks initiate and evolve, the AE counts and cumulative AE energy gradually increase. Notably, at the onset of macro-failure in the samples, both metrics rise rapidly with a steep slope. Therefore, the rapid growth of cumulative AE energy can be identified as an effective precursor of shale failure. It is worth noting that when the bedding dip angle $\theta = 30°$ and 60°, the AE counts and AE energy peak almost simultaneously. However, for $\theta = 0°$, the maximum AE energy lags behind the maximum AE counts, suggesting the formation of numerous non-penetrating cracks prior to macro-failure. Additionally, when $\theta = 90°$, a significant number of AE events continue to occur after macro-failure, indicating that the sample retains a relatively high residual strength.

4.5 Conclusion

In this paper, to investigate the impact of bedding surfaces on the anisotropic strength and deformation characteristics of shales under dynamic loading, a series of Brazilian disc tests were conducted using the SHPB system. The fracturing process and failure modes were closely monitored using a high-speed camera. Furthermore, to gain insights into the mechanisms of fracture initiation and propagation, stress buildup, stress shadow, and stress transfer were modeled based on digital image processing and the rock failure process analysis method. The key findings of this study are summarized as follows:

(1) At the inception of dynamic loading, several microcracks emerge at the weakest sections of the bedding planes. These initial cracks are often so minute, with minimal crack opening, that they are barely visible. Subsequently, due to the influence of the bedding structure, the cracking direction may alter, resulting in the formation of arc-shaped cracks. These cracks have the potential to coalesce and penetrate through adjacent bedding layers, ultimately culminating in macro-fracture. In other words, the shale discs continue to generate parallel cracks and curved cracks due to the high stresses concentrated at the tips of the initial cracks. Macro-intersecting cracks form as a result of the gradual progression and coalescence of these pre-existing cracks.

(2) The tensile strength of shale discs subjected to dynamic Brazilian splitting exhibits a U-shaped trend as the bedding angle increases. Initially, it decreases, and then it increases with the growth of the bedding angle. Concurrently, as the bedding angle θ rises, the fracture length within the rock matrix also increases. Moreover, the bedding angle θ significantly influences the cracking percentage, particularly when θ transitions from 0° to 30°. Beyond $\theta = 30°$, the cracking percentages of both the bedding planes and the rock matrix fluctuate. Notably, the fracture length and cracking percentage of the bedding planes share a similar trend with increasing bedding angle θ.

(3) The numerical simulation aligns well with the SHPB experiment in terms of fracture characteristics and failure modes. When the bedding dip angle θ is 30° and the bedding spacing S_{bp} is 1.5 mm, a main crack forms at the center of the sample, accompanied by a bending inflection point at the middle right part due to the inclined bedding surfaces, indicative of a mixed-mode fracture. Conversely, at $\theta = 60°$, the main fracture occurs at the center of the three discs with S_{bp} values of 1.5 mm, 2.75 mm, and 4 mm. However, numerous secondary cracks, induced by high stress concentrations at the inclined bedding surfaces, emerge within the central fractured zone. Additionally, when $\theta = 90°$, a prominent central splitting forms regardless of the bedding spacing distance.

(4) Generally, as the bedding dip angle increases, the cracking percentage of the bedding surfaces decreases while the cracking percentage of the rock matrix increases. Furthermore, the fracture force curve of Brazilian discs with varying bedding spacings exhibits a U-shaped trend with increasing bedding angle. Notably, at the inception of macro failure, the acoustic emission (AE) counts surge rapidly with a steep slope, indicating that the samples begin to generate a significant number of AE signals. Simultaneously, the cumulative energy attains its maximum value swiftly. Therefore, the acceleration in crack growth and the rapid increase in AE energy can be considered effective precursors of shale failure.

References

Aminzadeh, A., Fahimifar, A., Nejati, M.: On Brazilian disk test for mixed-mode I/II fracture toughness experiments of anisotropic rocks. Theoret. Appl. Fract. Mech. **102**, 222–238 (2019). https://doi.org/10.1016/j.tafmec.2019.04.010

Bahaaddini, M., Sharrock, G., Hebblewhite, B.K.: Numerical investigation of the effect of joint geometrical parameters on the mechanical properties of a non-persistent jointed rock mass under uniaxial compression. Comput. Geotech. **49**, 206–225 (2013). https://doi.org/10.1016/j.compgeo.2012.10.012

Brara, A., Camborde, F., Klepaczko, J.R., Mariotti, C.: Experimental and numerical study of concrete at high strain rates in tension. Mech. Mater. **33**(1), 33–45 (2001). https://doi.org/10.1016/S0167-6636(00)00035-1

Dai, F., Xu, Y., Zhao, T., Xu, N.W., Liu, Y.: Loading-rate-dependent progressive fracturing of cracked chevron-notched Brazilian disc specimens in split Hopkinson pressure bar tests. Int. J. Rock Mech. Min. Sci. **88**, 49–60 (2016). https://doi.org/10.1016/j.ijrmms.2016.07.003

Dan, D.Q., Konietzky, H., Herbst, M.: Brazilian tensile strength tests on some anisotropic rocks. Int. J. Rock Mech. Min. Sci. **58**, 1–7 (2013). https://doi.org/10.1016/j.ijrmms.2012.08.010

Duan, K., Kwok, C.Y., Pierce, M.: Discrete element method modeling of inherently anisotropic rocks under uniaxial compression loading. Int. J. Numer. Anal. Meth. Geomech. **40**, 1150–1183 (2016). https://doi.org/10.1002/nag.2476

Duan, K., Kwok, C.Y.: Discrete element modeling of anisotropic rock under Brazilian test conditions. Int. J. Rock Mech. Min. Sci. **78**, 46–56 (2015). https://doi.org/10.1016/j.ijrmms.2015.04.023

Eliyahu, M., Emmanuel, S., Day-Stirrat, R.J., Macaulay, C.I.: Mechanical properties of organic matter in shales mapped at the nanometer scale. Mar. Pet. Geol. **59**, 294–304 (2015). https://doi.org/10.1016/j.marpetgeo.2014.09.007

Fan, X., Luo, N., Liang, H., Sun, X., Zhai, C., Xie, L.: Dynamic breakage characteristics of shale with different bedding angles under the different ambient temperatures. Rock Mech. Rock Eng. **54**, 3245–3261 (2021). https://doi.org/10.1007/s00603-021-02463-6

Feng, X.H., Gong, B., Cheng, X.F., Zhang, H.H., Tang, C.A.: Anisotropy and microcrack-induced failure precursory of shales under dynamic splitting. Geomat. Nat. Haz. Risk **13**(1), 2864–2889 (2022)

Guha Roy, D., Singh, T.N.: Effect of heat treatment and layer orientation on the tensile strength of a crystalline rock under Brazilian test condition. Rock Mech. Rock Eng. **49**, 1663–1677 (2016). https://doi.org/10.1007/s00603-015-0891-y

Harris, N.B., Miskimins, J.L., Mnich, C.A.: Mechanical anisotropy in the Woodford Shale, Permian Basin: origin, magnitude, and scale. Leading Edge **30**, 284–291 (2011). https://doi.org/10.1190/1.3567259

Huang, B., Li, L., Tan, Y., Hu, R., Li, X.: Investigating the meso-mechanical anisotropy and fracture surface roughness of continental shale. J. Geophys. Res. Solid Earth **125**(8), e2019JB017828 (2020a). https://doi.org/10.1029/2019JB017828

Huang, D., Li, B., Ma, W.Z., Cen, D.F., Song, Y.X.: Effects of bedding planes on fracture behavior of sandstone under semicircular bending test. Theoret. Appl. Fract. Mech. **108**, 102625 (2020b). https://doi.org/10.1016/j.tafmec.2020.102625

Khanlari, G.R., Heidari, M., Sepahigero, A.A., Fereidooni, D.: Quantification of strength anisotropy of metamorphic rocks of the Hamedan province, Iran, as determined from cylindrical punch, point load and Brazilian tests. Eng. Geol. **169**, 80–90 (2014). https://doi.org/10.1016/j.enggeo.2013.11.014

Khanlari, G., Rafiei, B., Abdilor, Y.: An Experimental investigation of the Brazilian tensile strength and failure patterns of laminated sandstones. Rock Mech. Rock Eng. **48**, 843–852 (2015). https://doi.org/10.1007/s00603-014-0576-y

Li, A., Liu, Y., Dai, F., Liu, K., Wei, M.: Continuum analysis of the structurally controlled displacements for large-scale underground caverns in bedded rock masses. Tunn. Undergr. Space Technol. **97**, 103288 (2020). https://doi.org/10.1016/j.tust.2020.103288

Liang, Z.Z., Xing, H., Wang, S.Y., Williams, D.J., Tang, C.A.: A three-dimensional numerical investigation of the fracture of rock specimens containing a pre-existing surface flaw. Comput. Geotech. **45**, 19–33 (2012). https://doi.org/10.1016/j.compgeo.2012.04.011

Lifshitz, J.M., Leber, H.: Data processing in the split Hopkinson pressure bar tests. Int. J. Impact Eng **15**, 723–733 (1994). https://doi.org/10.1016/0734-743X(94)90011-9

Liu, Z.J., Zhang, C.Q., Zhang, C.S., Gao, Y., Zhou, H., Chang, Z.R.: Deformation and failure characteristics and fracture evolution of cryptocrystalline basalt. J. Rock Mech. Geotech. Eng. **11**, 990–1003 (2019). https://doi.org/10.1016/j.jrmge.2019.04.005

Ma, T., Chen, P.: Study of meso-damage characteristics of shale hydration based on CT scanning technology. Pet. Explor. Dev. **41**, 249–256 (2014). https://doi.org/10.1016/S1876-3804(14)60029-X

Ma, L., Wang, M., Zhang, N., Fan, P., Li, J.A.: Variable-parameter creep damage model incorporating the effects of loading frequency for rock salt and its application in a bedded storage cavern. Rock Mech. Rock Eng. **50**, 2495–2509 (2017). https://doi.org/10.1007/s00603-017-1236-9

Ma, T., Peng, N., Zhu, Z., Zhang, Q., Yang, C., Zhao, J.: Brazilian tensile strength of anisotropic rocks: review and new insights. Energies **11**, 204 (2018). https://doi.org/10.3390/en11020304

Meng, Y., Jing, H., Liu, X., Yin, Q., Wei, X.: Experimental and numerical investigation on the effects of bedding plane properties on the mechanical and acoustic emission characteristics of sandy mudstone. Eng. Fract. Mech. **245**, 107582 (2021). https://doi.org/10.1016/j.engfracmech.2021.107582

Qian, X.K., Liang, Z.Z., Liao, Z.Y., Wang, K.: Numerical investigation of dynamic fracture in rock specimens containing a pre-existing surface flaw with different dip angles. Eng. Fract. Mech. **223**, 106675 (2020). https://doi.org/10.1016/j.engfracmech.2019.106675

Shang, J., Hencher, S.R., West, L.J.: Tensile strength of geological discontinuities including incipient bedding, rock joints and mineral veins. Rock Mech. Rock Eng. **49**, 4213–4225 (2016). https://doi.org/10.1007/s00603-016-1041-x

Shang, J., Duan, K., Gui, Y., Handley, K., Zhao, Z.: Numerical investigation of the direct tensile behaviour of laminated and transversely isotropic rocks containing incipient bedding planes with different strengths. Comput. Geotech. **104**, 373–388 (2018). https://doi.org/10.1016/j.compgeo.2017.11.007

Sharafisafa, M., Aliabadian, Z., Shen, L.: Crack initiation and failure development in bimrocks using digital image correlation under dynamic load. Theoret. Appl. Fract. Mech. **109**, 102688 (2020). https://doi.org/10.1016/j.tafmec.2020.102688

Sun, Q., Zhang, Z., Peichao, L.I., Sun, Z., Detang, L.U., Gao, F.: Study on the bedding effect and damage constitutive model of black shale under dynamic loading. Chin. J. Rock Mech. Eng. **38**, 1319–1331 (2019)

Suo, Y., Chen, Z., Rahman, S.S., Song, H.: Experimental and numerical investigation of the effect of bedding layer orientation on fracture toughness of shale rocks. Rock Mech. Rock Eng. **53**, 3625–3635 (2020). https://doi.org/10.1007/s00603-020-02131-1

Tang, C.A., Liu, H., Lee, P.K.K., Tsui, Y., Tham, L.G.: Numerical studies of the influence of microstructure on rock failure in uniaxial compression—part I: effect of heterogeneity. Int. J. Rock Mech. Min. Sci. **37**, 555–569 (2000). https://doi.org/10.1016/S1365-1609(99)00121-5

Tien, Y.M., Kuo, M.C., Juang, C.H.: An experimental investigation of the failure mechanism of simulated transversely isotropic rocks. Int. J. Rock Mech. Min. Sci. **43**, 1163–1181 (2006). https://doi.org/10.1016/j.ijrmms.2006.03.011

Vervoort, A., Min, K.B., Konietzky, H., Cho, J.W., Debecker, B., Dinh, Q.D., Frühwirtc, T., Tavallalia, A.: Failure of transversely isotropic rock under Brazilian test conditions. Int. J. Rock Mech. Min. Sci. **70**, 343–352 (2014). https://doi.org/10.1016/j.ijrmms.2014.04.006

Wang, H., Liu, D., Cui, Z., Cheng, C., Jian, Z.: Investigation of the fracture modes of red sandstone using XFEM and acoustic emissions. Theoret. Appl. Fract. Mech. **85**, 283–293 (2016). https://doi.org/10.1016/j.tafmec.2016.03.012

Wang, P., Cai, M., Ren, F.: Anisotropy and directionality of tensile behaviours of a jointed rock mass subjected to numerical Brazilian tests. Tunn. Undergr. Space Technol. **73**, 139–153 (2018). https://doi.org/10.1016/j.tust.2017.12.018

Wang, D.J., Tang, H., Shen, P., Su, X., Huang, L.: Co-effects of bedding planes and parallel flaws on fracture evolution in anisotropic rocks. Eng. Geol. **264**, 105382 (2020a). https://doi.org/10.1016/j.enggeo.2019.105382

Wang, W., Zhao, Y., Teng, T., Zhang, C., Jiao, Z.: Influence of bedding planes on Mode I and mixed-Mode (I–II) dynamic fracture toughness of coal: analysis of experiments. Rock Mech. Rock Eng. **54**, 173–189 (2020b). https://doi.org/10.1007/s00603-020-02250-9

Wang, Z., Yang, S., Tang, Y.: Mechanical behavior of different sedimentary rocks in the Brazilian test. Bull. Eng. Geol. Env. **79**, 5415–5432 (2020c). https://doi.org/10.1007/s10064-020-01906-8

Wu, Z. Zuo, Y., Wang S., Yi, T., Chen, S., Yu, Q., Li, W., Sunwen, J., Xu., Y., Wang, R., Li, B.: Numerical simulation and fractal analysis of mesoscopic scale failure in shale using digital images. J. Petrol. Sci. Eng. **145**, 592–599 (2016). https://doi.org/10.1016/j.petrol.2016.06.036

Wu, Z., Lou, Y., Yin, S., Wang, A., Liu, H., Sun, W., Zuo, Y., Chen, B.: Acoustic and fractal analyses of the mechanical properties and fracture modes of bedding-containing shale under different seepage pressures. Enerfy Sci. Eng. **8**, 3638–3656 (2020). https://doi.org/10.1002/ese3.772

Xu, G., He, C., Chen, Z., Su, A.: Transverse isotropy of phyllite under brazilian tests: laboratory testing and numerical simulations. Rock Mech. Rock Eng. **51**, 1111–1135 (2018). https://doi.org/10.1007/s00603-017-1393-x

Xu, G., He, C., Yan, J., Ma, G.: A new transversely isotropic nonlinear creep model for layered phyllite and its application. Bull. Eng. Geol. Env. **78**, 5387–5408 (2019). https://doi.org/10.1007/s10064-019-01462-w

Yu, Z., Zhu, S., Wu, Y., Yu, H.: Study on the structural characteristics of the overburden under thick loose layer and thin-bed rock for safety of mining coal seam. Environ. Earth Sci. **79**, 9 (2019). https://doi.org/10.1007/s12665-019-8742-x

Zhang, Q.Y., Duan, K., Jiao, Y.Y., Xiang, W.: Physical model test and numerical simulation for the stability analysis of deep gas storage cavern group located in bedded rock salt formation. Int. J. Rock Mech. Min. Sci. **94**, 43–54 (2017). https://doi.org/10.1016/j.ijrmms.2017.02.015

Zhang, N., et al.: Analysis of mechanical and permeability properties of mudstone interlayers around a strategic petroleum reserve cavern in bedded rock salt. Int. J. Rock Mech. Min. Sci. **112**, 1–10 (2018). https://doi.org/10.1016/j.ijrmms.2018.10.014

Zhang, Z., et al.: Study on the mechanism of roof collapse and leakage of horizontal cavern in thinly bedded salt rocks. Environ. Earth Sci. **78**, 292 (2019). https://doi.org/10.1007/s12665-019-8292-2

Zhu, W.C., Tang, C.A.: Numerical simulation of Brazilian disk rock failure under static and dynamic loading. Int. J. Rock Mech. Min. Sci. **43**, 236–252 (2006). https://doi.org/10.1016/j.ijrmms.2005.06.008

Chapter 5
AE and Failure Behaviors of Tunnels by Strength Reduction Method

Abstract The investigation into the fracture features and failure mechanisms of tunnels excavated in both shallow and deep rock masses employs a series of three-dimensional heterogeneous models that incorporate various factors such as cross-section, tunnel alignment, faults, and tunnel support systems. The rock failure process analysis method, enhanced with the strength reduction method, is utilized to simulate the gradual fracture process, identify macro failure modes, determine safety factors, and replicate the characteristic fracture phenomena observed in surrounding rock masses. Furthermore, the mechanical mechanisms and acoustic emission energy of deep rocks at different stages of the entire zonal disintegration process are delved into. The findings reveal that the zonal disintegration process is initiated by stress redistribution. The cross-section of the tunnel plays a crucial role in influencing stress buildup, stress shadow, stress transfer, and the failure mode of the surrounding rock mass. Additionally, the dip of faults and the tunnel support system can alter the zonal disintegration pattern near the tunnel surface along their strikes. The presence of twin tunnels further complicates the zonal disintegration pattern. These insights contribute significantly to our understanding of zonal disintegration in deep engineering projects.

Keywords Tunnel failure · Strength reduction · Cross section · Faults · Support system · Tunnel layout

5.1 Introduction

As transportation networks expand into hilly regions, tunnel construction is experiencing rapid growth. Mountain tunnel construction often encounters complex geological conditions, including high ground stress, weak and fragmented surrounding rock, and well-developed geological structures (Zhang et al. 2016). In such environments, tunnel excavation can lead to significant and persistent deformation of the surrounding rock, which increases stress on supporting structures. If this stress exceeds the capacity of the surrounding rock and lining, initial support can fail,

© The Author(s), under exclusive license to Springer Nature Singapore Pte Ltd. 2025
X. Feng and B. Gong, *Mechanical Properties of Layered Rock Mass*,
https://doi.org/10.1007/978-981-96-5142-9_5

causing rock intrusion into the tunnel and potentially severe engineering disasters, posing significant threats to tunnel structure safety and constructors (Hu and Xu 2011). Therefore, a detailed geotechnical engineering analysis of failure mechanisms is crucial. Subsurface construction offers technical feasibility and reduces issues related to land acquisition and environmental impact, which are key considerations in today's development projects (Kovari 1994). Tunnel design must consider various engineering parameters, such as displacement stress distribution, properties of adjacent rock masses, faults, tunnel cross-section, and others (Deng et al. 2021; Hu et al. 2021). In deep tunnels with large overburden and poor rock mass characteristics, squeezing conditions and excessive loads around the excavation can compromise tunnel integrity, leading to extensive collapses (Chen et al. 2013; He et al. 2021). The fundamental challenge in deep excavation is understanding and simulating the rock failure process to develop an effective rock support system that ensures cost-effective and stable construction (Huang et al. 2011; Armand et al. 2014). Various research studies and laboratory tests have been conducted to investigate different design parameters in tunnel design (Lin et al. 2015; Zhang et al. 2017a, b; Qian and Zhou 2018). These studies have contributed to our understanding of tunnel behavior and have helped in the development of improved design and construction practices.

Wang et al. (2018) conducted a field test on 50 sections of a tunnel constructed using highly plastic clay layers, analyzing the deformation characteristics of adjacent rocks and the influencing factors such as burial depth, soft clay intrusion thickness, and atmospheric precipitation. Alexandr and Chi (cited in Ova 2004) deliberated on the impact of various tunnel cross-sectional shapes, including circles and horseshoes, on the Hanoi metro system's tunnels, concluding that circular tunnels are more suitable during earthquakes. Mobaraki and Vaghefi (2015) compared semi-ellipse, circular, and horseshoe-shaped tunnels and found that circular and horseshoe tunnels are less resistant to demolition compared to box-shaped tunnels. Yang et al. (2019) investigated the dynamic behavior of tunnel models featuring three distinct cross-sectional shapes—circle, rectangle, and horseshoe—and stated that cross-sectional shape primarily affects tunnel response in the nearfield, with this influence diminishing as the distance from the excitation source increases. Chen et al. (2022) revealed the fracture characteristics and mechanisms of circular tunnels excavated in deep jointed rock masses using a series of three-dimensional heterogeneous models that considered different joint dip angles.

Many researchers have examined the stability of underground tunnels and crack propagation from the perspective of tunnel cross-sectional shape. Chen et al. (2021) conducted biaxial compression tests to observe fracture pattern development and reported on the fracture phenomena. Lu et al. (2014) used the conformal mapping method for plane elasticity complex functions to analyze the optimal tunnel shape that meets specific optimization criteria. However, a research gap persists regarding the influence of cross-sectional shape on the stability of deep and shallow tunnels. As mining depth and stress environments become increasingly complex, a comparative study is needed to assess the impact of stress magnitude and loading rate on the stability of tunnels with different cross-sectional shapes (Bobet 2011; Kim et al. 2020). Therefore, this research endeavors to quantify the effects of cross-sectional

shape on crack propagation and tunnel failure processes by examining stress redistribution, displacement, geometric characteristics, and their impacts on crack propagation and damage development (Lin et al. 1984; Tang 1997; Jia and Zhu 2015; Gong et al. 2022).

Structural planes represent a pivotal geological risk factor that significantly impacts the stability of tunnel, slope, and dam constructions, markedly diminishing the strength of rock masses (Feng et al. 2024; Wang et al. 2022a, b). In tunnel excavation, fault rocks tend to expand stress relaxation zones, giving rise to ground stress instabilities that may culminate in collapses (Liu et al. 2007). Attributed to their low shear and tensile strengths, as well as the rock mass loosening caused by excavation unloading, these rock masses are prone to sliding along structural planes or detachment, bending, and fracture (Jiang et al. 2009; Ding et al. 2019). Occasionally, faults can provoke severe disasters in tunnel construction. To investigate tunnel deformation and failure characteristics, model testing and numerical analysis are commonly employed. Zhang et al. (2017a, b, 2020) conducted scaled model tests to explore the impact of faults and grouting on tunnel stability. Seo et al. (2016) conducted a parametric study using two-dimensional discontinuous deformation analysis to assess how joint attributes in a rock mass affect the stability of tunnels excavated within it. Manouchehrian and Ming (2018) analyzed the influence of specific fault parameters on tunnel stability.

In recent years, the development of underground spaces has encountered intricate geological conditions, particularly in squeezing ground areas, leading to numerous tunnel construction issues such as excessive deformation, support structure failure, and potential collapses (Mahabadi et al. 2012). Tunnel excavation necessitates the installation of a support system to ensure underground cavern stability by controlling rock mass displacement and addressing safety concerns (Luo et al. 2020). The selection and design of the support system are intricately linked to various factors in the overall tunnel design, aiming for functionality and cost-effectiveness. The support type, excavation method, and ground nature are interdependent. Poor route design or unsuitable construction techniques can render lining improvements economically insignificant (Zhang et al. 2017a, b). However, specific linings excel based on tunnel layout and construction method (Barton et al. 1995). Wu and Shao (2019) found that flexible support layers excel in absorbing deformations due to rock rheology compared to rigid systems in viscoelastic rocks. Examples of tunnel failures include the Zhegushan and Laodongshan tunnels in China (Zhang and Zhao 2015), Tunnel 35 of the Ankara-Istanbul high-speed railway project in Turkey, and the Kaligandaki tunnel in Nepal (Ghorasaini and Singh 2019). Tunneling in tight ground presents numerous technical challenges due to complex geology (Ko and Jeong 2017). A thorough understanding of the primary influencing factors and mechanisms of structural deterioration in challenging geological settings can significantly reduce tunnel excavation problems (Shaalan et al. 2019). Essential prerequisites for rational design include knowledge of support system requirements, effective performance criteria, familiarity with available system capabilities, and validated analysis methods. Future practice improvements hinge on a clear comprehension of current method limitations and needs (Hoek 1998; Chakeri et al. 2011; Simser 2019).

In this study, an extensive investigation is conducted into the fracture characteristics and failure mechanisms of tunnels excavated within deep rock formations. We utilize a comprehensive series of three-dimensional heterogeneous models that incorporate diverse parameters, including cross-section profiles, tunnel alignments, fault structures, and tunnel support systems. To simulate the gradual fracture process, macro failure modes, and safety factors, the rock failure process analysis (RFPA) method is integrated with the strength reduction method. This approach enables us to accurately replicate the distinct fracture phenomena commonly observed in deep rock masses. Furthermore, our research delves deeply into the mechanical mechanisms at play and the acoustic emission energy patterns exhibited by the surrounding rock formations during the various stages of zonal disintegration. This comprehensive analysis provides valuable insights into the intricate processes influencing the stability of deep rock tunnels.

5.2 Methodology

The planned Hongyancun tunnel and Xietaizi connection ramp project traverse the tectonic denudation hilly area situated between the two major water systems of the Yangtze River and Jialing River. The tunnel's surrounding rock is predominantly composed of sandy mudstone, characterized by a relatively intact block mosaic structure. The project site's landform is a structural denudation shallow hill, devoid of adverse geological phenomena such as landslides or faults, and exhibiting typical layer succession. Overall, the rock and soil mass at the site are stable. However, potential issues related to site and foundation stability arising from engineering construction primarily concern the stability of the underground tunnel cavern and the subgrade slope, which must be addressed in a well-thought-out design.

In order to solve the stability problem related to rock engineering structures, we combined the fundamental principles of strength reduction method into the RFPA method (Tang et al. 2020; Feng et al. 2022; Gao et al. 2023). RFPA has been applied in simulating rock failure by many researchers (Chen et al. 2023; Wang et al. 2023a, b; Yu et al. 2022). We also employed the acoustic emission (AE) event rate as the criterion for rock engineering failure. The main feature of the RFPA improved by strength reduction method for stability analysis of rock engineering is that they can obtain the factor of safety without any assumption for the shape and location of the failure surface. We applied the RFPA improved by strength reduction method to assess the stability of the surrounding rock in this study. We initially considered that all FEM elements were elastic, with their elastic properties defined by Young's modulus and Poisson's ratio. Each element's stress–strain curve was linearly elastic until it reached the chosen damage threshold. We used the maximum tensile strain (or stress) criterion and the Mohr–Coulomb criterion to set the damage criteria. In the tensile mode, we used the tensile strain criterion to determine whether the element was damaged. In the shear mode, if the element was not damaged in the tensile mode, we used the Mohr–Coulomb criterion to determine whether it was damaged.

The essential notion of strength reduction is introduced into the constitutive model of the element as an alternate solution to the failure analysis problem connected to geological or rock engineering. Each element is subjected to the shear strength reduction technique, and the constitutive model's strength f_0 is determined linearly as follows:

$$f_0^{trial} = \frac{f_0}{f_s^{trial}} \tag{5.1}$$

In the formula, F_s^{trial} is the trial safety factor and F_0^{trial} is the trial strength of the element. The trial strength F_0^{trial} will be used in RFPA to investigate the strength of the rock masses.

Two kinds of cavities with different section shapes are selected, i.e., circular cavity and horseshoe cavity. The section size and boundary condition of the cavity are shown in Fig. 5.1. Then, RFPA is used to perform numerical simulation with the plane strain model. We established numerical models with the size of 15 m × 15 m × 15 m in the X, Y and Z directions, respectively, and divided them into 100 × 100 × 100 = 100,000 units. In RFPA, the rock is assumed to be composed of many mesoscopic elements of the same size, and the mechanical properties of these elements are assumed to follow a given Weibull distribution (Gong et al. 2024).

For shallow tunnel, we applied vertical stress of 2.75 MPa on the upper part of the model, horizontal stress of 8 MPa on both sides, and fixed the bottom, respectively, i.e., models MC0 and MH0. For deep tunnel, the vertical stress is kept unchanged and applied horizontal stress of 4 MPa, i.e., models MC1 and MH1. The rock mass yield criterion is based on the Mohr–Coulomb criterion and the parameter selection of rock mass is shown in Table 5.1.

The strength reduction method is adopted in the simulation to keep the boundary stress unchanged, and the strength of rock mass and joint strength are continuously reduced with the reduction coefficient of 0.01) until the failure of instability (judging based on the maximum failure number of elementary elements). Thereafter, the safety factor can be obtained. Meanwhile, the influence of fault located above 1 m of tunnels on tunnel stability is analysed under four dip angles α of 0°, 30°, 60° and 90°. Four conditions of lateral pressure coefficient (0.8, 1.0, 1.2 and 1.4) are also considered. Besides, the vertical stress of 2.75 MPa is applied on the upper part of the model, the horizontal stress of 8 MPa is on both sides, and the bottom is fixed. The size and boundary conditions of the model are shown in Fig. 5.1c, and the parameter selection of rock mass and joints during simulation is shown in Table 5.2.

Additionally, the numerical analysis of the failure mechanism of the rigid and flexible tunnel support systems is carried out. As shown in Fig. 5.1e and f, the support system of circular tunnel is analysed according to the plane strain problem. A numerical model with plain concrete as a rigid support and steel fiber reinforced (SFR) concrete as flexible support is created as shown in Fig. 5.1d.

The thickness of support lining considered in this simulation is 0.3 m. The calculation area is divided into 100 × 100 × 100 quadrilateral iso parametric units. The static loading of the numerical model is applied to simulate the deep ground stress

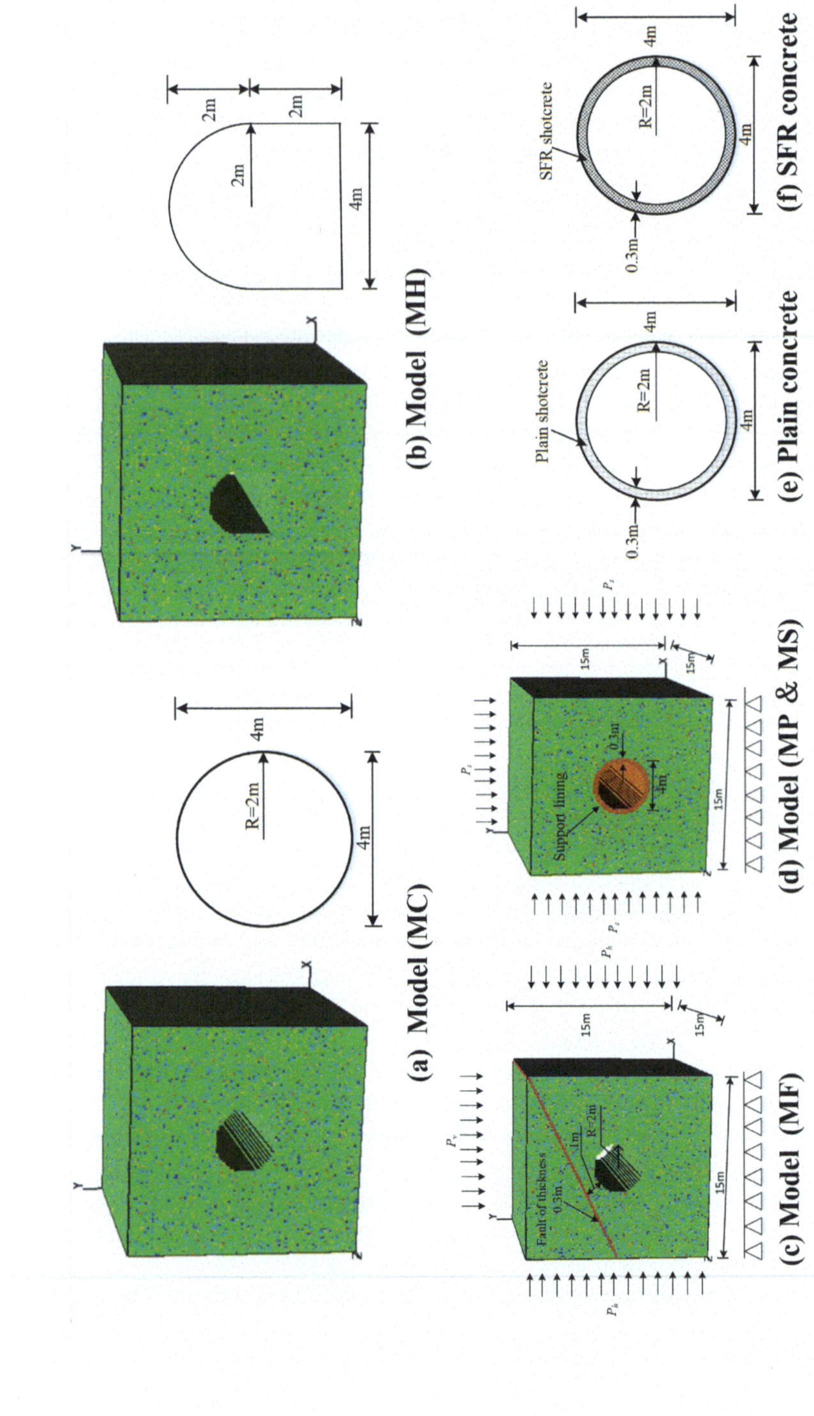

Fig. 5.1 Configuration of cross section and numerical models. Modified after Feng et al. (2025)

Table 5.1 Physical and mechanical parameters of rock (Feng et al. 2025)

Material	E_0 (MPa)	C (MPa)	Poisson ratio	Friction angle (°)	Density (kg/m^3)
Rock	70	60	0.3	30	2400

Table 5.2 Physical and mechanical parameters of fault (Feng et al. 2025)

Material	E_0 (MPa)	C (MPa)	Poisson ratio	Friction angle (°)	Density (kg/m^3)
Fault	70	30	0.3	30	2400

Table 5.3 Physical and mechanical parameters of model material (Feng et al. 2025)

Material	E_0 (MPa)	C (MPa)	Poisson ratio	Friction angle (°)	Density (kg/m^3)
Rigid	31	80	0.3	30	2400
Flexible	41	200	0.25	30	2400

environment of the rock surrounding the laneway. The bottom of the model is fixed, and the same static load of 8 MPa is applied on both sides of the model and the vertical load of 2.75 MPa is adopted on the top of the model. The average self-weight of the overburden strata of the model is assumed to be 25 kN/m^3. The mechanical parameters of the units are assumed to be assigned according to a Weibull distribution, and the values of the physical and mechanical parameters are shown in Table 5.3. Three different models of horizontal, inclined and vertical alignment are devised, and the circular tunnel with a radius of 2000 mm is excavated in different alignment as shown in Fig. 5.2.

5.3 Results and Discussion

5.3.1 Section Optimization Effect

(a) Failure process of tunnels with circular cavity

The maximum principal stress diagram and its corresponding acoustic emission diagram of tunnels with circular cavity are shown in Fig. 5.3. The failure process of circular cavity in shallow tunnel are similar to the top of horseshoe where roof damage precedes the floor damage and is more severe than the latter. Discontinuous spalling occurred at the mid height of the side wall of the circular tunnel after excavation, and micro cracks in the surrounding rock were initially generated from the mid height of the side wall at the 66th reduction step. Under the increasing external load, the cracks then extended concentrically to greater depth in the rock mass surrounding the tunnel. At 74th step, the zonal disintegration becomes obviously and continuously, the cracks penetrated on the deep masses of surrounding rock. The cracks of the sidewall,

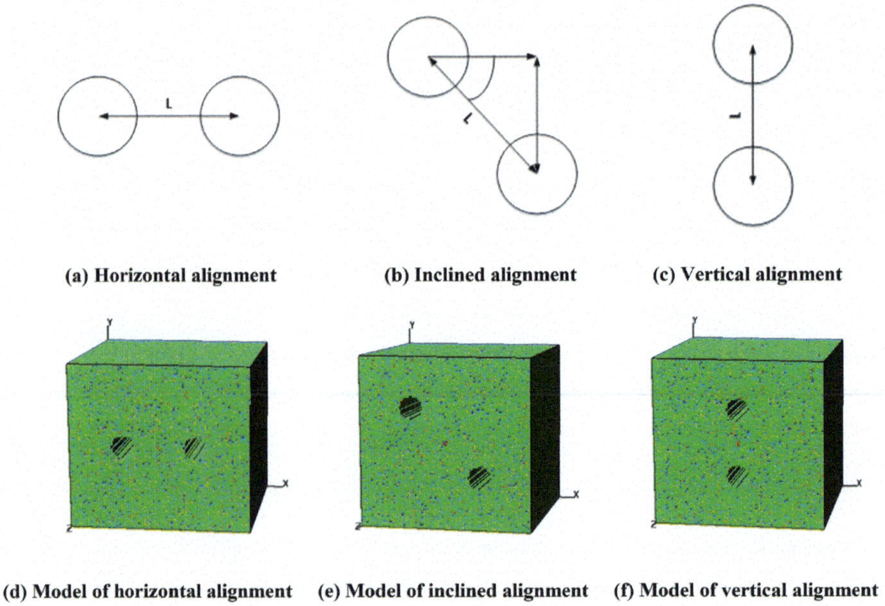

Fig. 5.2 Configuration and numerical model of twin tunnels. Modified after Feng et al. (2025)

bottom and roof are completely detached from the surrounding rock resulting in the tunnel losing its stability.

The maximum principal stress diagram and its corresponding acoustic emission diagram in the failure process of circular cavity in deep tunnel are shown in Fig. 5.3. Due to the excavation under high overburden stresses in deep tunnelling, the distribution of the zonal disintegration tends to appear on all the periphery of model due to the rock overstressing. The failure process is somehow similar to the shallow tunnel, it also shows the continuous damage evolution under the loading state. At 73rd step, the concentration of the stress tends to appear at both the sidewall, the zonal disintegration in the left and right sidewall of tunnel seems begin simultaneously. When the strength of the rock mass continues weakening and at 77-1st step, the cracks in the left sidewall developed into deeper surrounding rock, the disintegration on the roof and bottom is not prominent. Finally, at 78-2rd step, the side wall, the vault and invert of the tunnel collapsed completely loses its stability.

(b) Failure process of tunnels with horseshoe cavity

For the shallow tunnel with horseshoe cavity in Fig. 5.4, continuous spalling was found to occur at the foot of the tunnel and micro cracks in the surrounding rock were initially generated from the foot of the side wall at 42rd step. Under the increasing external load, the cracks developed upwards to form a conjugate sliding shape at the foot of the arch roof, where the cracks finally coalesced. At 59th reduction step, cracks are initially generated at the two right sides of the tunnel base, the vault is

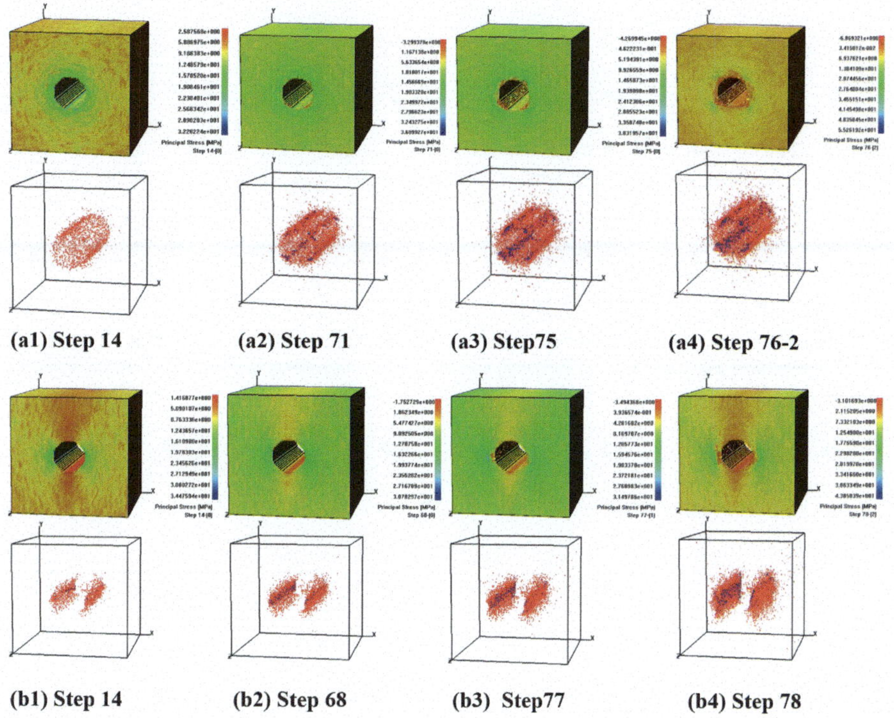

Fig. 5.3 Shear stress and AE diagram of shallow (series (a)) and deep (series (b)) circular tunnels. Modified after Feng et al. (2025)

next damaged. However, the cracks at the top of horseshoe shaped tunnels generate and propagate to form smaller triangular zonal disintegration. Similarly, with the further reduction of the surrounding rock strength, the stress transfers to the deeper part of the further reduction of the surrounding rock again. When the strength of the rock mass continues weakening and at 75-7th step, the roof and floor of the tunnel happen obvious damages, the right bottom corner displays the detachment. The tunnel detaches from the surrounding rock and completely loses its stability.

For deep tunnel, the zonal disintegration with horseshoe cavity in Fig. 5.4 seems to appear early than the shallow tunnel. Unlike the shallow tunnel, the concentration of stress tends to appear at floor and the bottom of the tunnel. Then the cracks extend to the center of the roof and floor of the tunnel. Then zonal disintegration started from the cavity wall are mainly on the left sides along the joint extension. Under the increasing external load, the cracks also develop on the left side wall of the tunnel to some extent. The zonal disintegration on the left and right sidewall shows the huge damages, which is highly consistent with the classical rib spalling during tunnel excavation in Fig. 5.5. The middle part of the floor also happened the ridgeline damages but unlike the shallow tunnel, the roof of the tunnel does not happen damages. Finally, considering that the zonal fractures of surrounding rocks

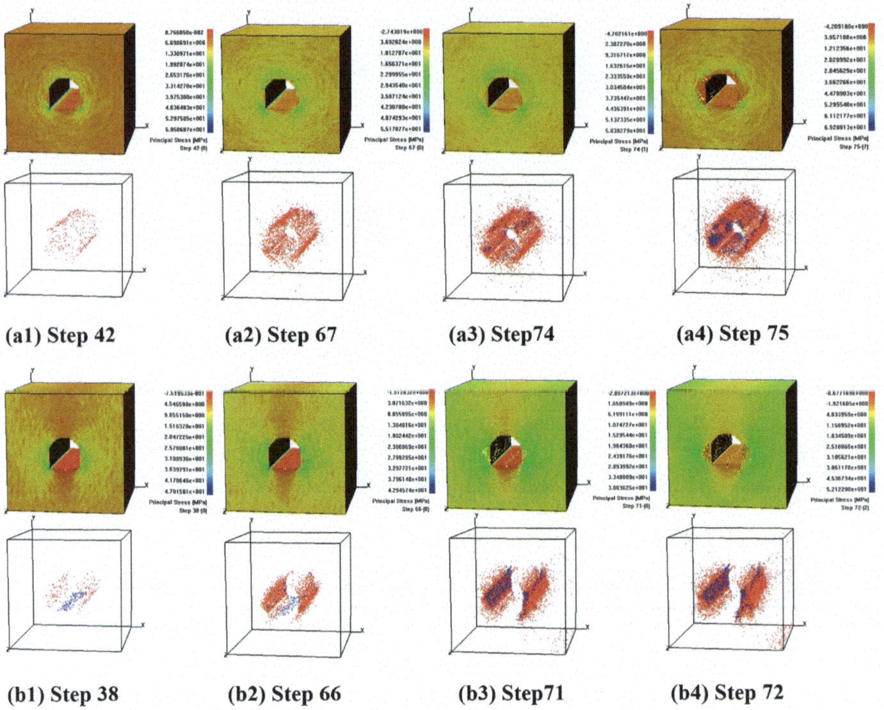

Fig. 5.4 Shear stress and AE diagram of shallow (series (**a**)) and deep (series (**b**)) horseshoe cavity. Modified after Feng et al. (2025)

generally happen in high-stress environments in deep engineering, which may result in unimaginable accidents, it is necessary to further study more effective support methods.

(c) AE distribution and safety factor

The results of AE counts and AE accumulated energy in tunnels with different sections from excavation to instability in shallow and deep tunnel are shown in Fig. 5.6. Meanwhile, the calculated safety factors of the different cross-section tunnels are as follows: 4.00 for the model M_{C0}, 2.38 for the model M_{H0}, 1.44 for the

Fig. 5.5 The classical rib spalling during tunnel excavation. Modified after Feng et al. (2025)

ount of damage due to the heterogeneity model. When the ıd reach to 86-7th step, many of the units in the rigid roof are royed. The damages on all the four corners of vault appears very ntinues damages evolution to the 86–23 reduction step, Under al load, the roof of the laneway undergoes large deformation and ı sides of the laneway will also be destroyed, the rigid support ɔse its supporting ability, there is an example of slabbing during ion as in Fig. 5.10.

sions of flexible and rigid supports subject to loading are shown .9b1–b4 show that the laneway undergoes shear failure under ıal load. Unlike the rigid support, the element damages in the quite less. Under the increasing external load, only a few units xible support damage under static loading. When the damages o 98–11 reduction step, the damages in the bottom and top left ıre obvious but the damage on the sidewall and roof of the tunnel ent. However, the overall laneway is stable, no roof collapse and enomenon is observed, and the steel fiber reinforced support are hening the carrying capacity of the surrounding rock. The stress

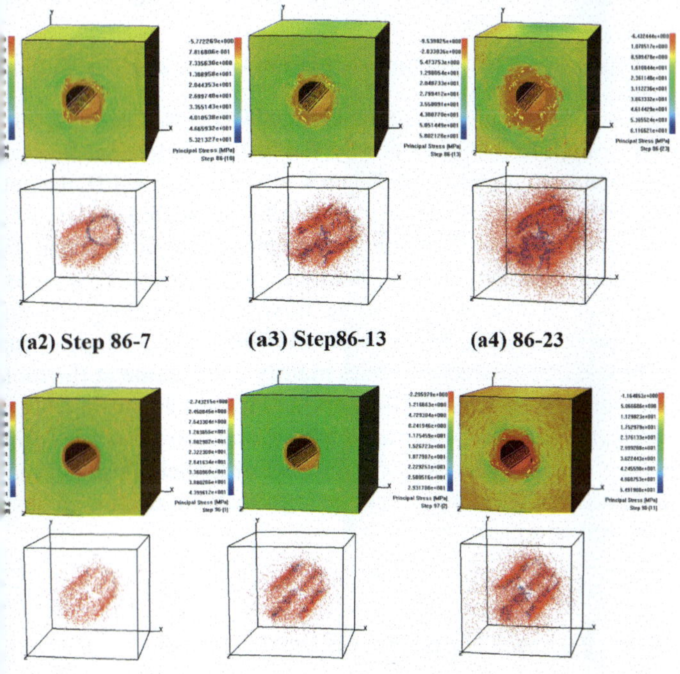

(a2) Step 86-7 (a3) Step86-13 (a4) 86-23

(b2) Step 96-1 (b3) Step97-2 (b4) Step 98-11

agram and acoustic emission distribution of rigid tunnel support (series (**a**)) support (series (**b**)). Modified after Feng et al. (2025)

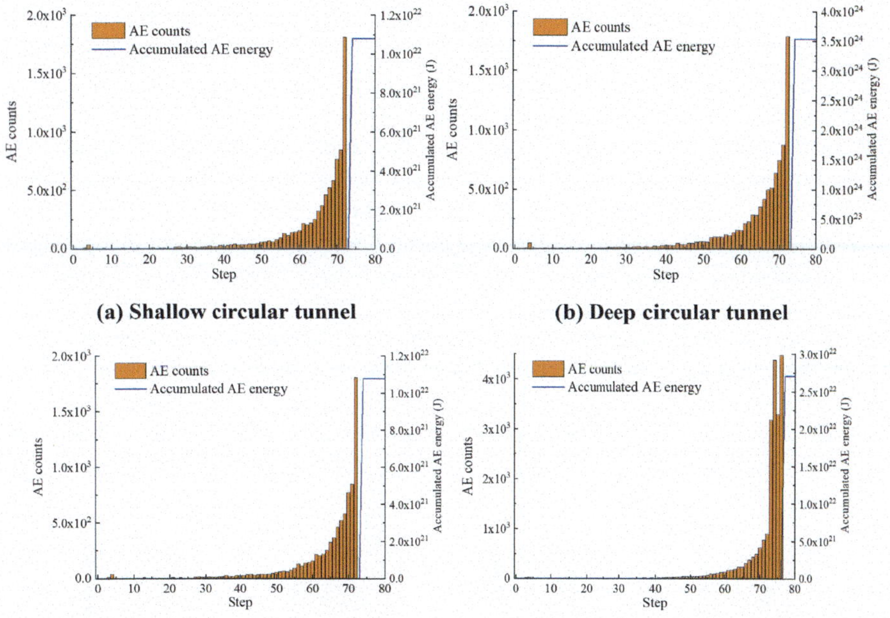

(a) Shallow circular tunnel **(b) Deep circular tunnel**

(c) Shallow horseshoe tunnel **(d) Deep horseshoe tunnel**

Fig. 5.6 AE results of circular and horseshoe cavities. Modified after Feng et al. (2025)

model M_{C1} and 3.00 for the model M_{H1}. We can see that as many as 2000 acoustic emission events occur before macro instability, and there is significant up toward trend before researching the overall failure of the tunnel, demonstrating a salient failure precursory characteristic of the inhomogeneous surrounding rock. Besides, acoustic emission energy is released by the damaged elements. However, the quantity and distribution continuity of AE and the frequency characteristics of accumulated energy are quite different. The circular cavity tunnel has the better continuity of AE distribution than horseshoe shaped tunnels. This shows that horseshoe tunnels have the worse stability and are more vulnerable to damage; while circular tunnels have the better stability, because they loss stability later than the horseshoe tunnels.

5.3.2 *Influence of Existing Faults*

(a) Effect of dip angles on the stability of tunnel

Fault located at different locations can lead to different influences on the deformation, the stress and zonal disintegration surrounding rock mass. Figure 5.7 shows the stress distribution of surrounding rock under different fault dip angles and their corresponding acoustic emission of failure state, it can be seen that during the initial excavation, the surrounding rock stress is concentrated at the arch waist, and the stress shows symmetry. Under the increasing external load, the surrounding rock is damaged at the arch waist, and the stress concentration area also moves to the tip of the propagation crack. At the same time, there is also a high stress area at the fault, and finally a through crack appears between the tunnel and the fault. It can also be seen from the figure that the stress concentration degree of surrounding rock near the fault side is significantly higher than that on the without fault side.

When $\alpha = 0°$, the stress first appears in the arch waist and shows symmetry, which leads to damage and then extends to both sides; with the increase of fault dip angle, the stress distribution changes from the symmetrical distribution of intact rock mass and close to the fault stress migration. From the acoustic emission distribution in Fig. 5.8 a1–a4, we can understand that the fault has great influence on the transmission of stress, strain and strain energy. When the tunnel is excavated, the rock mass between the fault and the tunnel will bear greater stress and store more energy, resulting in large deformation of the rock mass in this area, collapse or even rock burst at the

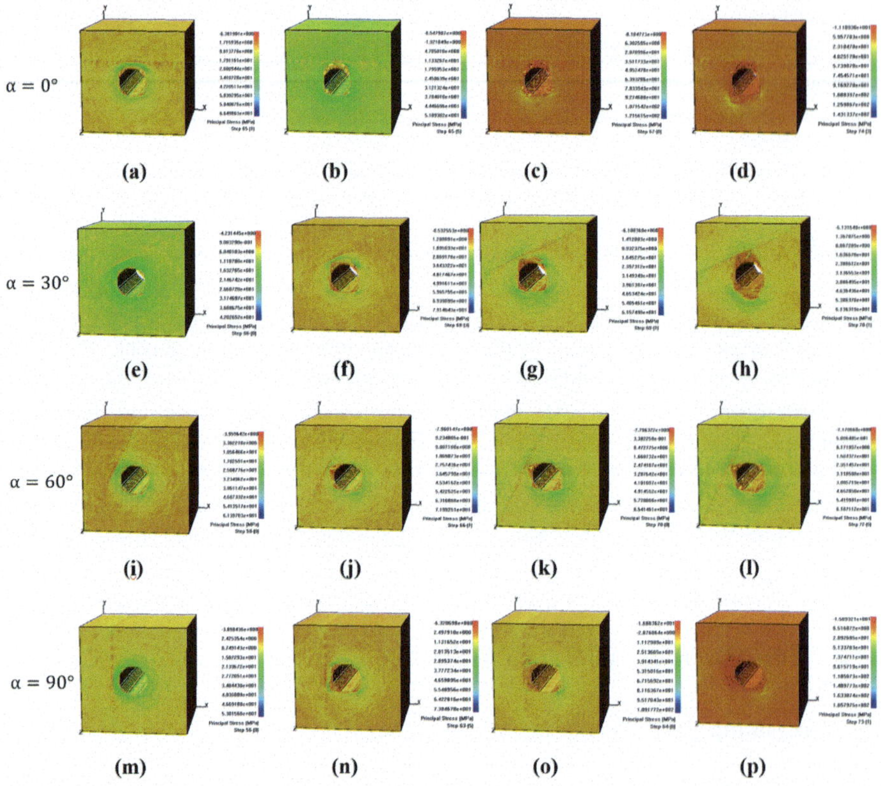

Fig. 5.7 Shear stress diagram with different dip angles of fault: **a** Step 65–1; **b** Step 65–5; **c** Step 67; **d** Step 74–3; **e** Step 66; **f** Step 68–3; **g** Step 69-1; **h** Step 70–1; **i** Step 58; **j** Step 66–1; **k** Step 70; **l** Step 72–5; **m** Step 56; **n** Step 63–5; **o** Step 64; **p** Step 73–1. Modified after Feng et al. (2025)

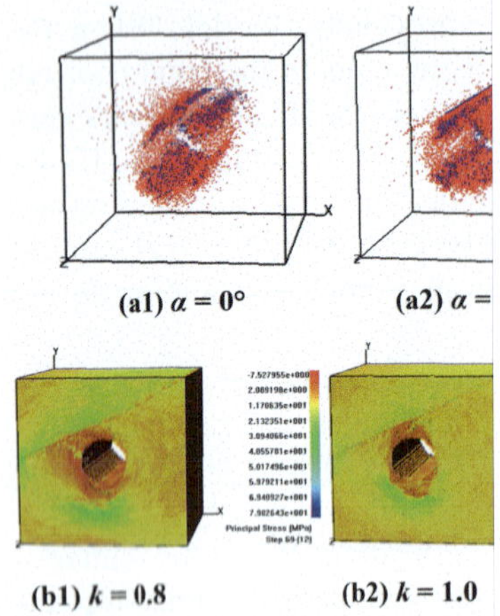

Fig. 5.8 Acoustic emission distributi
stress contours. Modified after Feng et

tunnel boundary can occur, and t
cause the relative sliding of the

(b) Effect of lateral pressure co

A series of lateral pressure coeffi
with dip angle of 30°, respectivel
are shown in Fig. 5.8b1–b4. Wh
confined in the regions near the le
bottom of left sidewall and the j
only occur at roof of tunnels, an
failure pattern of tunnel is spalli
the shape of fractured zones on

For k < 1.2, the fracture zone
where the rock mass cut by fault
k = 1.2, the fracture zones occu
the fracture zones only occur at
the failure mode of tunnel, and t
the failure of tunnel.

5.3.3 Tunnel Support Sy

The numerical simulation result
rock with plain shotcrete as rigid
a4. Under the increasing extern

undergo a smal
damages contin
instantaneously
obviously. On th
the increasing ex
destruction, and
has been thrust a
circle tunnel exc

The acoustic e
in Fig. 5.9. Figu
the increasing ex
flexible support
in the roof of the
continue and rea
corner of the tun
are not very pro
laneway spalling
not damaged, stre

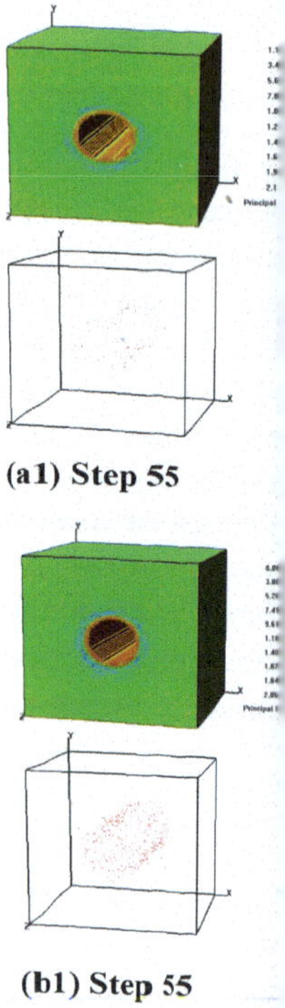

Fig. 5.9 Shear stress
and SFR flexible tun

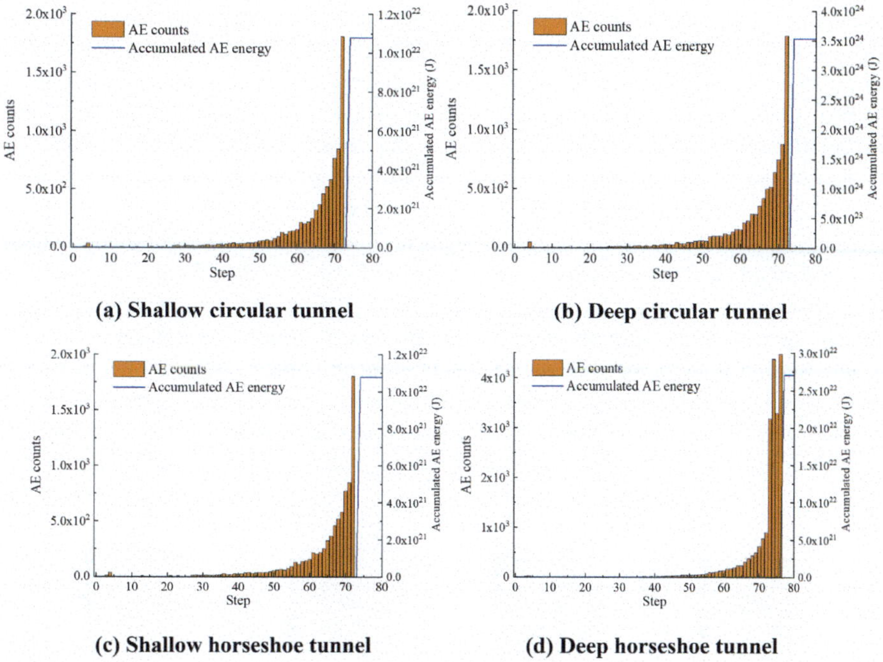

Fig. 5.6 AE results of circular and horseshoe cavities. Modified after Feng et al. (2025)

model M_{C1} and 3.00 for the model M_{H1}. We can see that as many as 2000 acoustic emission events occur before macro instability, and there is significant up toward trend before researching the overall failure of the tunnel, demonstrating a salient failure precursory characteristic of the inhomogeneous surrounding rock. Besides, acoustic emission energy is released by the damaged elements. However, the quantity and distribution continuity of AE and the frequency characteristics of accumulated energy are quite different. The circular cavity tunnel has the better continuity of AE distribution than horseshoe shaped tunnels. This shows that horseshoe tunnels have the worse stability and are more vulnerable to damage; while circular tunnels have the better stability, because they loss stability later than the horseshoe tunnels.

5.3.2 *Influence of Existing Faults*

(a) Effect of dip angles on the stability of tunnel

Fault located at different locations can lead to different influences on the deformation, the stress and zonal disintegration surrounding rock mass. Figure 5.7 shows the stress distribution of surrounding rock under different fault dip angles and their corresponding acoustic emission of failure state, it can be seen that during the initial excavation, the surrounding rock stress is concentrated at the arch waist, and the stress shows symmetry. Under the increasing external load, the surrounding rock is damaged at the arch waist, and the stress concentration area also moves to the tip of the propagation crack. At the same time, there is also a high stress area at the fault, and finally a through crack appears between the tunnel and the fault. It can also be seen from the figure that the stress concentration degree of surrounding rock near the fault side is significantly higher than that on the without fault side.

When $\alpha = 0°$, the stress first appears in the arch waist and shows symmetry, which leads to damage and then extends to both sides; with the increase of fault dip angle, the stress distribution changes from the symmetrical distribution of intact rock mass and close to the fault stress migration. From the acoustic emission distribution in Fig. 5.8 a1–a4, we can understand that the fault has great influence on the transmission of stress, strain and strain energy. When the tunnel is excavated, the rock mass between the fault and the tunnel will bear greater stress and store more energy, resulting in large deformation of the rock mass in this area, collapse or even rock burst at the

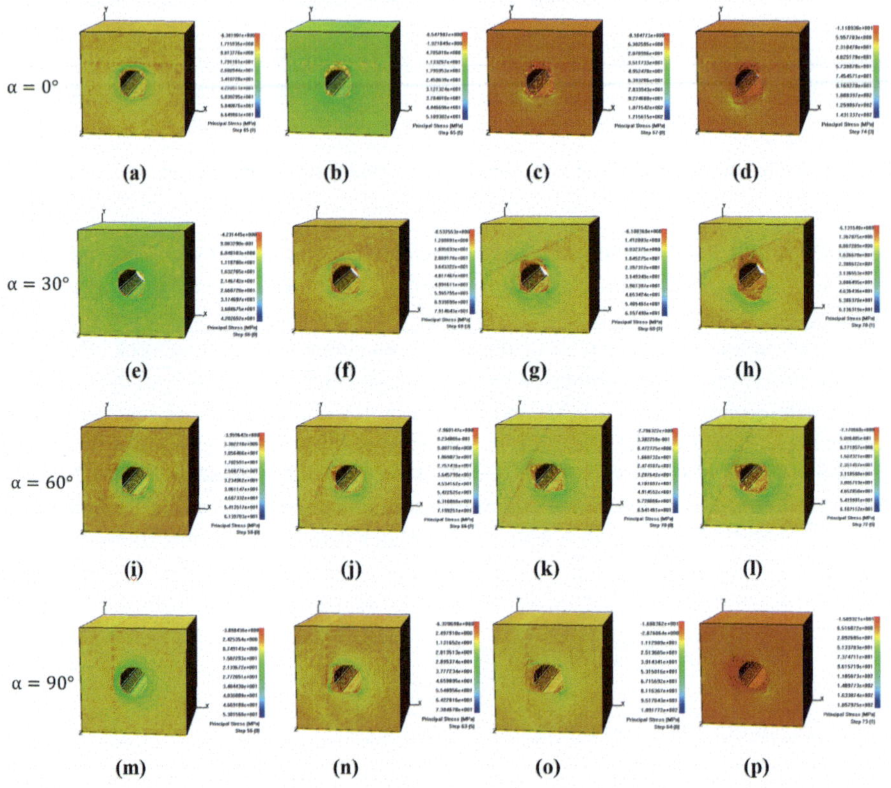

Fig. 5.7 Shear stress diagram with different dip angles of fault: **a** Step 65–1; **b** Step 65–5; **c** Step 67; **d** Step 74–3; **e** Step 66; **f** Step 68–3; **g** Step 69-1; **h** Step 70–1; **i** Step 58; **j** Step 66–1; **k** Step 70; **l** Step 72–5; **m** Step 56; **n** Step 63–5; **o** Step 64; **p** Step 73–1. Modified after Feng et al. (2025)

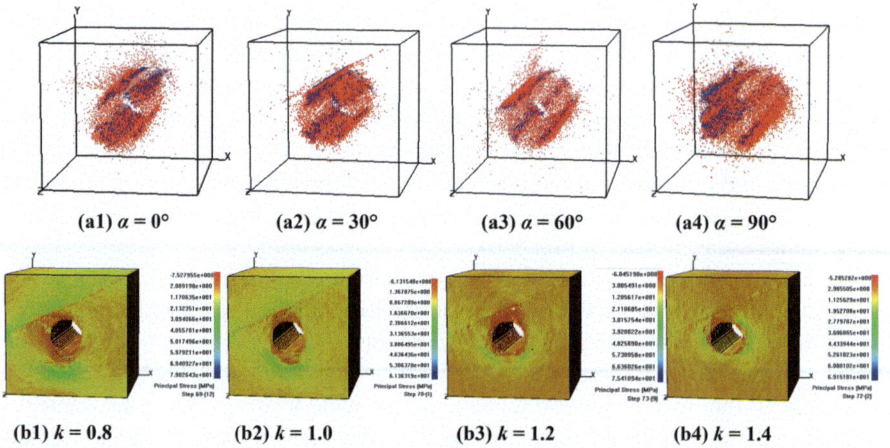

Fig. 5.8 Acoustic emission distribution with different dip angles of fault (series (**a**)) and shear stress contours. Modified after Feng et al. (2025)

tunnel boundary can occur, and the crack extends to the depth, which may eventually cause the relative sliding of the upper and lower walls of the fault.

(b) Effect of lateral pressure coefficient on the stability of tunnel

A series of lateral pressure coefficient of 0.8, 1.0, 1.2 and 1.4 are applied on models with dip angle of 30°, respectively, the failure patterns at four typical reduction steps are shown in Fig. 5.8b1–b4. When $k = 0.8$ and $k = 1.0$, the detachment of joints is confined in the regions near the left shoulder of tunnel. At $k = 1.2$, cracks initiate at the bottom of left sidewall and the jointed place of right sidewall. For $k = 1.4$, fractures only occur at roof of tunnels, and there are no fracture zones at both sidewalls. The failure pattern of tunnel is spalling, flexing and breaking of tunnel rock mass, which the shape of fractured zones on the roof take on the shape of triangle.

For $k < 1.2$, the fracture zones mainly concentrate at the roof and floor of tunnel, where the rock mass cut by faults begins to detach, flex and break into the tunnel; for $k = 1.2$, the fracture zones occur at roof and two sidewalls of the tunnel; for $k > 1.2$, the fracture zones only occur at the roof of tunnel. The faults have obvious effect on the failure mode of tunnel, and the high tectonic stress will have dominate effect on the failure of tunnel.

5.3.3 Tunnel Support System

The numerical simulation results of the failure process of the laneway surrounding rock with plain shotcrete as rigid support under static loading are shown in Fig. 5.9a1–a4. Under the increasing external load, some weaker units in the rigid support first undergo a small amount of damage due to the heterogeneity model. When the damages continue and reach to 86-7th step, many of the units in the rigid roof are instantaneously destroyed. The damages on all the four corners of vault appears very obviously. On the continues damages evolution to the 86–23 reduction step, Under the increasing external load, the roof of the laneway undergoes large deformation and destruction, and both sides of the laneway will also be destroyed, the rigid support has been thrust and lose its supporting ability, there is an example of slabbing during circle tunnel excavation as in Fig. 5.10.

The acoustic emissions of flexible and rigid supports subject to loading are shown in Fig. 5.9. Figure 5.9b1–b4 show that the laneway undergoes shear failure under the increasing external load. Unlike the rigid support, the element damages in the flexible support are quite less. Under the increasing external load, only a few units in the roof of the flexible support damage under static loading. When the damages continue and reach to 98–11 reduction step, the damages in the bottom and top left corner of the tunnel are obvious but the damage on the sidewall and roof of the tunnel are not very prominent. However, the overall laneway is stable, no roof collapse and laneway spalling phenomenon is observed, and the steel fiber reinforced support are not damaged, strengthening the carrying capacity of the surrounding rock. The stress

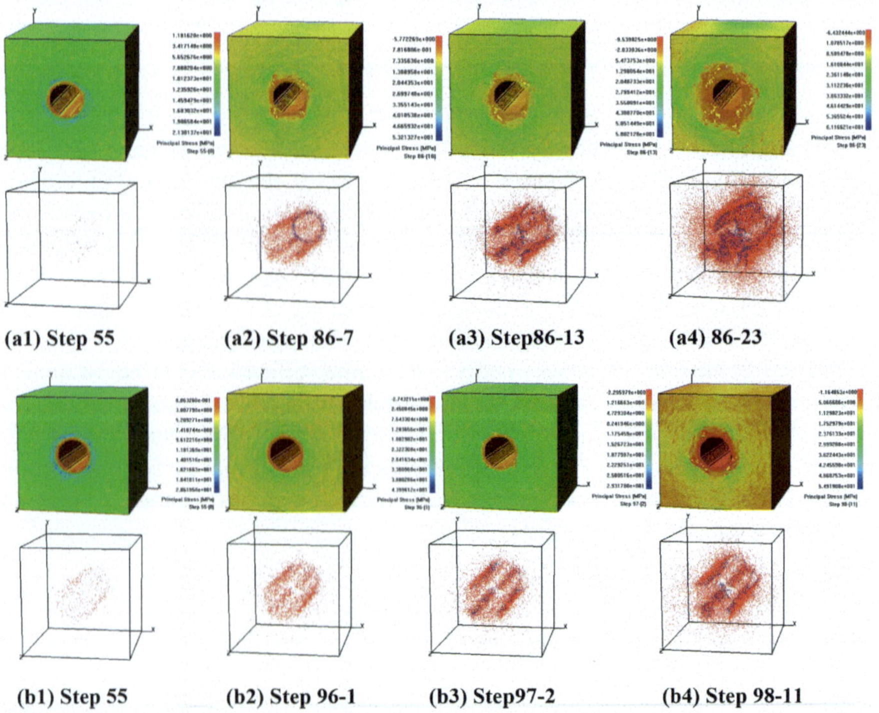

Fig. 5.9 Shear stress diagram and acoustic emission distribution of rigid tunnel support (series (**a**)) and SFR flexible tunnel support (series (**b**)). Modified after Feng et al. (2025)

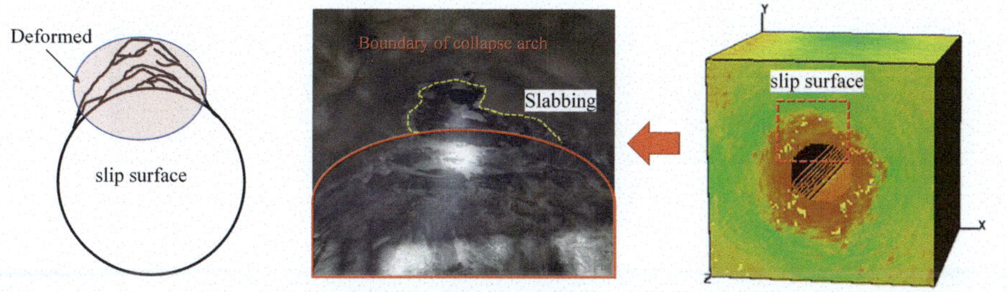

Fig. 5.10 An example of slabbing during circle tunnel excavation. Modified after Feng et al. (2025)

in the failure area of the steel fiber reinforced flexible bearing is tensile stress. When the step load continues to increase, the fracture develops, and the flexible support plays its tensile property. With the absorption energy of surrounding rock deformation and axial tensile deformation of flexible support, the deformation energy of tunnel can be relieved and released, so as to avoid the instantaneous fracture caused by the further reduction of the rock strength.

5.3.4 Influence of Tunnel Alignment

(a) Tunnels with horizontal alignment

The maximum principal stress diagram and corresponding acoustic emission diagram in the failure process with horizontal alignment tunnel are shown in Fig. 5.11a1–a4. With the rock mass strength weakening under the triaxial stress state, from which can be observed that the first fractured zone adjacent to the tunnel wall appear at the 39th reduction step. Then the stress and damage from the cavity wall on both sides along the joint extension, left and right sides of the expansion rate were similar. When the damages continue and reach to 81st reduction step, the damage in both side of tunnel begins to appear. When the compressive / tensile strength continues reducing to the 87th step, the rock mass at a certain distance away from the first fracture starts to show the obvious damages. When it reaches to 87–19 reduction step, the tunnel lose stability from the bottom and along the horizontal line of the two tunnels.

(b) Tunnels with inclined alignment

The shear stress and AE diagram of inclined aligned tunnel are shown in Fig. 5.11b1–b4. The strength of the rock mass contuses weakening and when reaches to 85 reduction step, the damage in both the tunnel becomes obvious. However, upper left tunnel shows the damage mainly on the right wall and the lower right tunnel shows the damage on the and left wall. When it reaches to 85th reduction step, the bottom of both the tunnel shows the obvious damage, the damage on the lower tunnel is more prominent, the roof wall started to detach from the surrounding rock. At the

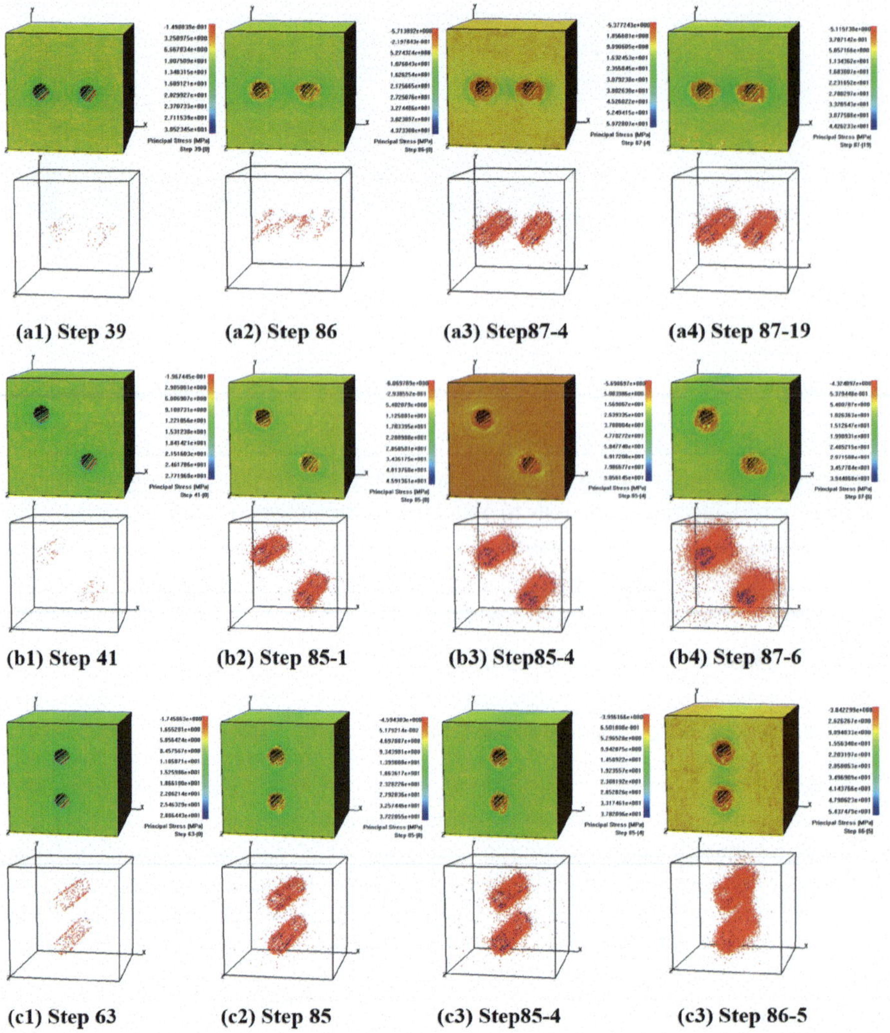

Fig. 5.11 Shear stress and AE diagram of horizontally, vertically and vertically aligned tunnel. Modified after Feng et al. (2025)

reduction step 87–6, the tunnel completely loses its stability and along with the radial damages around cavity, the damage seems to be more concentrated diagonally and around the cavity.

(c) Tunnels with vertical alignment

The shear stress and AE diagram of vertically aligned tunnel are shown in Fig. 5.11c1–c4. When it reaches to 85–4 reduction step, the damage on the lower tunnel is more severe than the upper tunnel, the floor of the tunnel completely detaches, and the vault seems collapsed. Compared with the horizontal tunnel, the zonal fracture modes of vertical aligned tunnel show similarities to a certain extent. Consequently, the rapture area of the fracture ring located above the tunnel develops along the vertical line of the tunnel.

5.4 Conclusion

To understand the fracture features and reveal the failure mechanisms of circle tunnels excavated in deep rock masses, a series of three-dimensional heterogeneous models considering considers the effects of the tunnel cross section, tunnel spacing, tunnel alignment, support system, dip angle of faults and the lateral pressure coefficient of faulted rock mass on the stability of the tunnel are established. The strength reduction method is embedded to achieve the gradual fracture process, final failure mode and safety factor and to reproduce the characteristic fracture phenomenon of tunnels under high geo-stress level, i.e., zonal disintegration. The following conclusions could be drawn:

(1) The circle cross-section is considered to be the optimal cross-section due the force around the vault and the arch bottom is symmetrical, the safety factor of horseshoe is large. The horseshoe cross-section should be avoided in the tunnel design, because the safety factor of horseshoe cross-section is small, and the phenomena is same in the case of deep tunnel.
(2) The fault has obvious influence on the stress field and displacement field around the tunnel, and it has obvious significance on the transmission of stress and displacement. Due to obvious obstruction, the rock mass between the fault and the tunnel bears high stress, and the tunnel cavity is easy to be damaged near the fault, and the degree of damage is severe.
(3) The alignment of tunnel has great influence on its stability, the zonal fracture modes of alignment tunnel show similarities to a certain extent. On the whole, the rapture area of the fracture ring located above the tunnel develops along the line of the tunnel.
(4) The examination of stress distribution, fracture progression, and cumulative acoustic emissions reveals distinct patterns in the behavior of the tunnel support systems. In the case of the rigid support, significant deformation and eventual collapse of the roof were observed, alongside noticeable damage to the two tunnel sidewalls. In contrast, the flexible support structure demonstrated minimal and controllable deformation, resulting in an overall stable condition of the laneway. These findings highlight the superior impact resistance of the flexible support system compared to its rigid counterpart.

References

Armand, G., Leveau, F., Nussbaum, C., de La Vaissiere, R., Noiret, A., Jaeggi, D., Landrein, P., Righini, C.: Geometry and properties of the excavation-induced fractures at the Meuse/Haute-Marne URL drifts. Rock Mech. Rock Eng. **47**(1), 21–41 (2014)

Barton, N., Grimstad, E., Palmstrom, A.: Design of tunnel support. In: Austin, S.A., Robins, P.J. (eds.) Chapter of Sprayed Concrete: Properties, Design and Application. Whittles Publishing, Dunbeath, UK (1995)

Bobet, A.: Lined circular tunnels in elastic transversely anisotropic rock at depth. Rock Mech. Rock Eng. **44**(2), 149–167 (2011)

Chakeri, H., Hasanpour, R., Hindistan, M.A., Ünver, B.: Analysis of interaction between tunnels in soft ground by 3D numerical modeling. Bull. Eng. Geol. Env. **70**(3), 439–448 (2011)

Chen, X., Wang, Y., Zhang, Q., Li, S., Nordlund, E.: Analogical model test and theoretical analysis on zonal disintegration based on filed monitoring in deep tunnel. Eur. J. Environ. Civ. Eng. **17**, s33–s52 (2013)

Chen, S., Guo, H., Su, J., Tang, D.: Study on the evaluation of surrounding rock stability and the criterion of limit displacement of fractured stratum tunnel. IOP Conf. Ser. Earth Environ. Sci. **643**, 012148 (2021)

Chen, B., Gong, B., Wang, S., Tang, C.: Research on zonal disintegration characteristics and failure mechanisms of deep tunnel in jointed rock mass with strength reduction method. Mathematics **10**, 922 (2022)

Chen, T., Gong, B., Tang, C.: Origin and evolution of cracks in the glaze surface of a ceramic during the cooling process. Materials **16**, 5508 (2023)

Deng, P., Liu, Q., Huang, X., Liu, Q., Ma, H., Li, W.: Acquisition of normal contact stiffness and its influence on rock crack propagation for the combined finite-discrete element method (FDEM). Eng. Fract. Mech. **242**, 107459 (2021)

Ding, X., Niu, X., Pei, Q., Huang, S., Zhang, Y., Zhang, C.: Stability of large underground caverns excavated in layered rock masses with steep dip angles: a case study. Bull. Eng. Geol. Env. **78**(7), 5101–5133 (2019)

Feng, X., Gong, B., Cheng, X., Zhang, H., Tang, C.: Anisotropy and microcrack-induced failure precursor of shales under dynamic splitting. Geomat. Nat. Haz. Risk **13**(1), 2864–2889 (2022)

Feng, X., Gong, B., Liang, Z., Wang, S., Tang, C., Li, H., Ma, T.: Study of the dynamic failure characteristics of anisotropic shales under impact Brazilian splitting. Rock Mech. Rock Eng. **57**, 2213–2230 (2024)

Feng, X., Prasad, Y.K., Gong, B., Cheng, X., Duan, W., Liu, H., Ren, F.: Investigation on the AE characteristics and progressive failure behaviors of tunnels by strength reduction method. Front. Earth Sci. **13**, 1426226 (2025)

Gao, M., Gong, B., Liang, Z., Jia, S., Zou, J.: Investigation of the anisotropic mechanical response of layered shales. Energy Sci. Eng. **11**(12), 4737–4754 (2023)

Ghorasaini, D., Singh, A.B.: Geotechnical engineering design of a tunnel support system—a case study of Bheri Babai headrace tunnel. Int. J. Eng. Trends Technol. **67**(4), 33–37 (2019)

Gong, B., Liang, Z.Z., Liu, X.: Nonlinear deformation and failure characteristics of horseshoe-shaped tunnel under varying principal stress direction. Arab. J. Geosci. **15**(6), 475 (2022)

Gong, B., Zhao, T., Thusyanthan, I., Tang, C.: Modelling rock fracturing by a novel implicit continuous to discontinuous method. Comput. Geotech. **166**, 106035 (2024)

He, S., Lai, J., Zhang, Y., Wang, K., Xu, W., Wang, L., Liu, T., Zhang, C.: Damage behaviors, prediction methods and prevention methods of rockburst in 13 deep traffic tunnels in China. Eng. Fail. Anal. **121**, 105178 (2021)

Hoek, E.: Tunnel support in weak rock. In: Keynote Address, Symposium of Sedimentary Rock Engineering, Taipei, Taiwan, 20–22 Nov 1998

Hu, J., Xu, N.: Numerical analysis of failure mechanism of tunnel under different confining pressure. Proc. Eng. **26**(1), 107–112 (2011)

Hu, Y., Wang, Q., Wang, M., Liu, D.: A study on the thermo-mechanical properties of shotcrete structure in a tunnel, excavated in granite at nearly 90 °C temperature. Tunn. Undergr. Space Technol. **110**, 103830 (2021)

Huang, R., Huang, D., Duan, S., Wu, Q.: Geomechanics mechanism and characteristics of surrounding rock mass deformation failure in construction phase for underground powerhouse of Jinping I hydropower station. Chin. J. Rock Mech. Eng. **30**(1), 23–35 (2011)

Jia, P., Zhu, W.C.: Mechanism of zonal disintegration around deep underground excavations under triaxial stress—insight from numerical test. Tunn. Undergr. Space Technol. **48**, 1–10 (2015)

Jiang, Y., Li, B., Yamashita, Y.: Simulation of cracking near a large underground cavern in a discontinuous rock mass using the expanded distinct element method. Int. J. Rock Mech. Min. Sci. **46**(1), 97–106 (2009)

Ko, J., Jeong, S.: A study on rock mass classifications and tunnel support systems in unconsolidated sedimentary Rock. Sustainability **9**, 573 (2017)

Kim, K., Rutqvist, J., Birkholzer, J.: Lattice modeling of excavation damage in argillaceous clay formations: Influence of deformation and strength anisotropy. Tunn. Undergr. Space Technol. **98**, 103196 (2020)

Kovari, K.: Erroneous concepts behind the New Austrian Tunnelling Method. Int. J. Rock Mech. Min. Sci. Geomech.s Abs. **32**(4), 188A (1994)

Lin, Y.J., Paddock, E.F., King, E.L.: Meiotic chromosome configurations in a triploid Rhoeo. Genetica **63**(2), 113–119 (1984)

Lin, P., Liu, H., Zhou, W.: Experimental study on failure behaviour of deep tunnels under high in-situ stresses. Tunn. Undergr. Space Technol. **46**, 28–45 (2015)

Liu, H.; Small, J.C.; Carter, J.P.: Three-dimensional modelling for effects of tunnelling on existing support systems. In: Proceedings for International Symposium on Numerical Models in Geomechanics, pp. 373–379, 25 Apr 2007. Rhodes, Greece.

Lu, A.Z., Chen, H.Y., Qin, Y., Zhang, N.: Shape optimisation of the support section of a tunnel at great depths. Comput. Geotech. **61**(1), 190–197 (2014)

Luo, Y., Chen, J., Shi, Z., Li, J.: Mechanical characteristics of primary support of large span loess highway tunnel: a case study in Shaanxi Province, Loess Plateau, NW China primary. Tunn. Undergr. Space Technol. **104**(1), 103532 (2020)

Mahabadi, O.K., Lisjak, A., Munjiza, A., Grasselli, G.: Y-Geo: New combined finite-discrete element numerical code for geomechanical applications. Int. J. Geomech. **12**(6), 676–688 (2012)

Manouchehrian, A., Ming, C.: Numerical modeling of rockburst near fault zones in deep tunnels. Tunn. Undergr. Space Technol. **80**, 164–180 (2018)

Mobaraki, B., Vaghefi, M.: Numerical study of the depth and cross-sectional shape of tunnel under surface explosion. Tunn. Undergr. Space Technol. Incorp. Trench. Technol. Res. **47**, 114–122 (2015)

Ova, O.G.: Quality in tunnelling: ITA-AITES working group 16 final report. Tunn. Undergr. Space Technol. **19**(3), 239–272 (2004)

Qian, Q., Zhou, X.: Failure behaviors and rock deformation during excavation of underground cavern group for Jinping I hydropower station. Rock Mech. Rock Eng. **51**(8), 2639–2651 (2018)

Seo, Y.S., Yun, H.S., Ban, J.D., Lee, C.K.: Mechanical properties of fault rocks in Korea. J. Eng. Geol. **26**(4), 571–581 (2016)

Shaalan, H.H., Ismail, M.A.M., Azit, R.: Effect of tunnel overburden stress on the rock brittle failure depth. Arab. J. Geosci. **12**(4), 108 (2019)

Simser, B.P.: Rockburst management in Canadian hard rock mines. J. Rock Mech. Geotech. Eng. **11**(5), 1036–1043 (2019)

Tang, C.: Numerical simulation of progressive rock failure and associated seismicity. Int. J. Rock Mech. Min. Sci. **34**(2), 249–261 (1997)

Tang, C.A., Webb, A.A.G., Moore, W.B., Wang, Y.Y., Ma, T.H., Chen, T.T.: Breaking Earth's shell into a global plate network. Nat. Commun. **11**(1), 3621 (2020)

Wang, G., Yuan, M., Miao, Y., Wu, J., Wang, Y.: Experimental study on seismic response of underground tunnel-soil-surface structure interaction system. Tunn. Undergr. Space Technol. **76**, 145–159 (2018)

Wang, Y.Y., Gong, B., Tang, C.: Numerical investigation on anisotropy and shape effect of mechanical properties of columnar jointed basalts containing transverse joints. Rock Mech. Rock Eng. **55**, 7191–7222 (2022a)

Wang, Y.Y., Gong, B., Zhang, Y., Yang, X., Tang, C.: Progressive fracture behavior and acoustic emission release of CJBs affected by joint distance ratio. Mathematics **10**, 4149 (2022b)

Wang, Y.Y., Gong, B., Tang, C.A., Yang, X.Y.: Size effect and lateral pressure effect on the mechanical resistance of columnar jointed basalt. Int. J. Rock Mech. Min. Sci. **171**, 105571 (2023a)

Wang, Y.Y., Gong, B., Yang, X.Y., Tang, C.A.: Investigation into the multistage mechanical damage behavior of columnar jointed basalts with different meso-constitutive relations and model sizes. Lithosphere **1**, 8711959 (2023b)

Wu, K., Shao, Z.: Study on the effect of flexible layer on support structures of tunnel excavated in viscoelastic rocks. J. Eng. Mech. **145**(10), 04019077 (2019)

Yang, W., Zhang, C., Liu, D., Tu, J., Yan, Q., Fang, Y., He, C.: The effect of cross-sectional shape on the dynamic response of tunnels under train induced vibration loads. Tunn. Undergr. Space Technol. **90**, 231–238 (2019)

Yu, C.Y., Gong, B., Wu, N., Xu, P.L., Bao, X.K.: Simulation of the fracturing process of inclusions embedded in rock matrix under compression. Appl. Sci. **12**(16), 8041 (2022)

Zhang, Q.B., Zhao, J.: A review of dynamic experimental techniques and mechanical behaviour of rock materials. Rock Mech. Rock Eng. **47**, 1411–1478 (2015)

Zhang, C., Liu, N., Chu, W.: Key technologies and risk management of deep tunnel construction at Jinping II hydropower station. J. Rock Mech. Geotech. Eng. **8**, 499–512 (2016)

Zhang, Q., Zhang, X., Wang, Z., Xiang, W., Xue, J.: Failure mechanism and numerical simulation of zonal disintegration around a deep tunnel under high stress. Int. J. Rock Mech. Min. Sci. **93**, 344–355 (2017a)

Zhang, Z., Chen, F., Li, N., Swoboda, G., Liu, H.: Influence of fault on the surrounding rock stability of a tunnel: location and thickness. Tunn. Undergr. Space Technol. Incorp. Trench. Technol. Res. **61**, 1–11 (2017b)

Zhang, Y., Zhao, C., Jiang, M., Zhang, J., Chen, C., Zheng, X., Sun, X., Xu, H.: Constant-resistance, rigid, and flexible coupling support technology for soft rock entrances in deep coal mines: a case study in China. Adv. Mater. Sci. Eng. **2020**, 6212456 (2020)

The manufacturer's authorised representative in the EU is Springer Nature Customer Service Centre GmbH, Europaplatz 3, 69115 Heidelberg, Germany. If you have any concerns regarding our products, please contact ProductSafety@springernature.com

Printed and bound by CPI Group (UK) Ltd, Croydon, CR0 4YY
01/10/2026
02225945-0003